BOOK 4

NAME:

CLASS:

NELSON MATHS

WORKBOOK

YEARS 7–10

Megan Boltze
Robert Yen
Ilhea Yen

PUZZLE SHEETS

Nelson Maths Workbook 4
1st Edition
Megan Boltze
Ilhea Yen
Robert Yen
ISBN 9780170454551

Publisher: Robert Yen
Project editor: Alan Stewart
Editor: Anna Pang
Cover design: James Steer
Original text design by Alba Design, Adapated by: James Steer
Project designer: James Steer
Permissions researcher: Corrina Gilbert
Production controller: Karen Young
Text illustrations: Cat MacInnes
Typeset by: MPS Limited

For product information and technology assistance,
in Australia call **1300 790 853**;
in New Zealand call **0800 449 725**

For permission to use material from this text or product, please email
aust.permissions@cengage.com

ISBN 978 0 17 045455 1

Cengage Learning Australia
Level 7, 80 Dorcas Street
South Melbourne, Victoria Australia 3205

Cengage Learning New Zealand
Unit 4B Rosedale Office Park
331 Rosedale Road, Albany, North Shore 0632, NZ

For learning solutions, visit **cengage.com.au**

Printed in China by 1010 Printing International Limited.
1 2 3 4 5 6 7 25 24 23 22 21

PREFACE

This 200-page workbook contains worksheets, puzzles, StartUp assignments and homework assignments written for the Australian Curriculum in Mathematics. It can be used as a valuable resource for teaching Year 10 mathematics, regardless of the textbook used in the classroom, and takes a holistic approach to the curriculum, including some Year 9 revision work as well. It also contains **4 bonus 10A chapters for extension**, especially for students preparing for Year 11 (see page v). This workbook is designed to be handy for homework, assessment, practice, revision, relief classes or 'catch-up' lessons.

Inside:

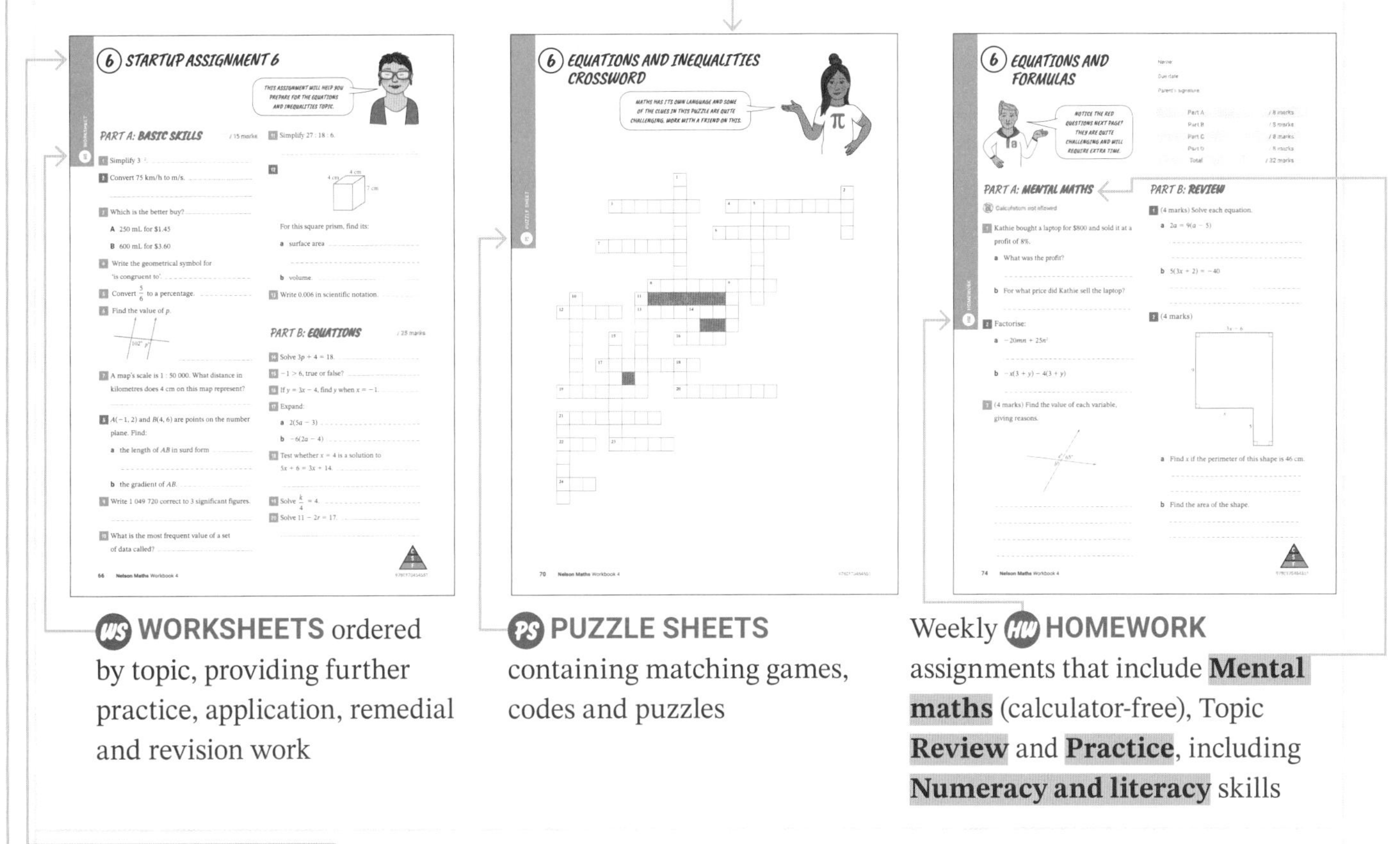

WS WORKSHEETS ordered by topic, providing further practice, application, remedial and revision work

PS PUZZLE SHEETS containing matching games, codes and puzzles

Weekly HW HOMEWORK assignments that include **Mental maths** (calculator-free), Topic **Review** and **Practice**, including **Numeracy and literacy** skills

StartUp assignments beginning each topic, revising skills from previous topics and prerequisite knowledge for the topic, including basic skills, review of a specific topic and a challenge problem

Word puzzles, such as a crossword or find-a-word, that reinforce the language of mathematics learned in the topic

The ideas and activities presented in this book were written by practising teachers and used successfully in the classroom.

Colour-coding of selected questions

Questions on most worksheets are graded by level of difficulty:

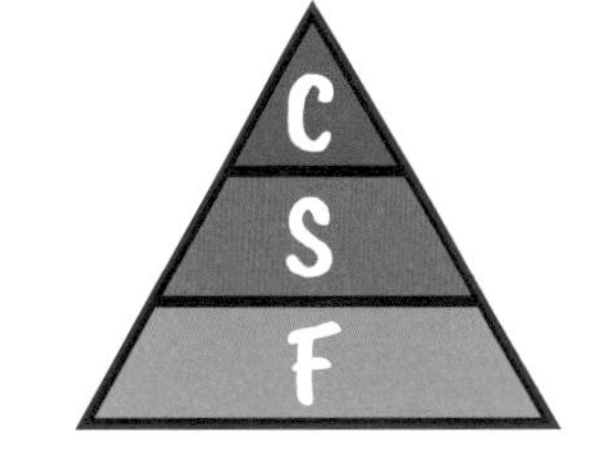

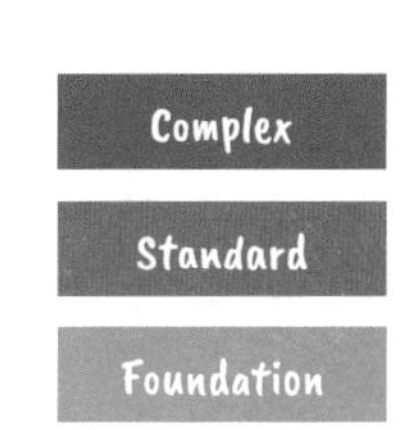

CONTENTS

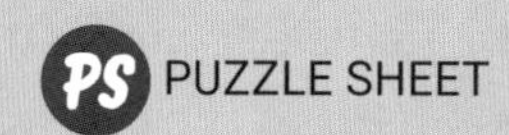

PREFACE III
CURRICULUM GRID VIII

1 INTEREST AND DEPRECIATION 1
- WS StartUp assignment 1 1
- WS Compound interest 3
- WS Depreciation 5
- PS Interest and depreciation crossword 7
- HW Compound interest 9

2 GRAPHING LINES 11
- WS StartUp assignment 2 11
- PS Linear equations code puzzle 14
- WS Writing equations of lines 16
- PS Graphing lines crossword 17
- HW Coordinate geometry 18
- HW Graphing lines 20

3 SURFACE AREA AND VOLUME 23
- WS StartUp assignment 3 23
- WS Back-to-front problems 26
- PS Surface area and volume crossword 28
- HW Surface area 30
- WS Surface area 33
- HW Volume 34

4 ALGEBRA 37
- WS StartUp assignment 4 37
- WS Algebraic fractions 39
- PS Algebra crossword 40
- HW Index laws 42
- HW Algebraic fractions 44
- HW Expanding and factorising 46
- HW Binomial products 48
- WS Mixed expansions 50

5 COMPARING DATA 51
- WS StartUp assignment 5 51
- WS Interquartile range 53
- WS Box-and-whisker plots 54
- PS Data crossword 56
- HW Data 58
- HW Boxplots 60
- HW Comparing data 62
- HW Scatterplots 64

6 EQUATIONS AND INEQUALITIES 66
- WS StartUp assignment 6 66
- WS Graphing inequalities 68
- PS Equations and inequalities crossword 70
- HW Equations 72
- HW Equations and formulas 74
- HW Inequalities 76
- WS Equation review 78

7 GRAPHING CURVES 79
- WS StartUp assignment 7 79
- WS Graphing parabolas 81
- WS Graphing exponentials 82
- WS Graphing hyperbolas 83
- PS Graphing curves crossword 84
- HW Graphing curves 86

CONTENTS

8 TRIGONOMETRY 88

- WS StartUp assignment 8 88
- WS Finding an unknown angle 90
- WS Elevations and bearings 92
- PS Trigonometry crossword 94
- HW Trigonometry 1 96
- HW Trigonometry 2 98
- WS A page of bearings 101

9 SIMULTANEOUS EQUATIONS 102

- WS StartUp assignment 9 102
- WS Intersection of lines 104
- PS Simultaneous equations crossword 105
- HW Simultaneous equations 1 106
- HW Simultaneous equations 2 109

10 PROBABILITY 111

- WS StartUp assignment 10 111
- PS Probability crossword 113
- WS Tree diagrams 114
- WS Two-way tables 116
- HW Probability 1 118
- HW Probability 2 120

11 GEOMETRY 122

- WS StartUp assignment 11 122
- WS Proving properties of quadrilaterals 124
- PS Geometry crossword 126
- HW Congruent figures 128
- HW Similar figures 130
- WS Geometrical proofs order activity 132

BONUS 10A CHAPTERS

12 PRODUCTS AND FACTORS 133

- WS StartUp assignment 12 133
- WS Special products 136
- PS Factorising puzzle 138
- HW Special binomial products 140

13 SURDS 142

- WS StartUp assignment 13 142
- WS Surds 144
- PS Simplifying surds 146
- WS Rationalising the denominator 147
- HW Surds 1 148
- HW Surds 2 150
- PS Surds crossword 152

14 QUADRATIC EQUATIONS AND THE PARABOLA 153

- WS StartUp assignment 14 153
- WS Graphing parabolas 2 155
- PS Quadratic equations puzzle 156
- PS Quadratic equations crossword 158
- HW Quadratic equations 160
- HW The parabola 162

15 FURTHER TRIGONOMETRY 164

- WS StartUp assignment 15 164
- WS Finding an unknown angle 167
- HW Further trigonometry 1 169
- HW Further trigonometry 2 172
- HW Further trigonometry 3 174

ANSWERS 177
HOMEWORK ANSWERS 191

MEET YOUR MATHS GUIDES ...

INTRODUCING MS LEE.

THIS WORKBOOK CONTAINS WORKSHEETS, PUZZLE SHEETS AND HOMEWORK ASSIGNMENTS

HI, I'VE BEEN TEACHING MATHS FOR OVER 20 YEARS

I BECAME GOOD AT MATHS THROUGH PRACTICE AND EFFORT

I WILL GUIDE YOU THROUGH THE WORKSHEETS

MATHS IS ABOUT MASTERING A COLLECTION OF SKILLS, AND I CAN HELP YOU DO THIS

THIS IS ZINA, A MATHS TUTOR AND MS LEE'S YEAR 12 STUDENT

HEY, I LOVE CREATING AND SOLVING PUZZLES

NOT JUST MATHS PUZZLES BUT WORD PUZZLES TOO!

PUZZLES HELP YOU THINK IN NEW AND DIFFERENT WAYS

LET ME SHOW YOU HOW, AND YOU'LL GET SMARTER ALONG THE WAY

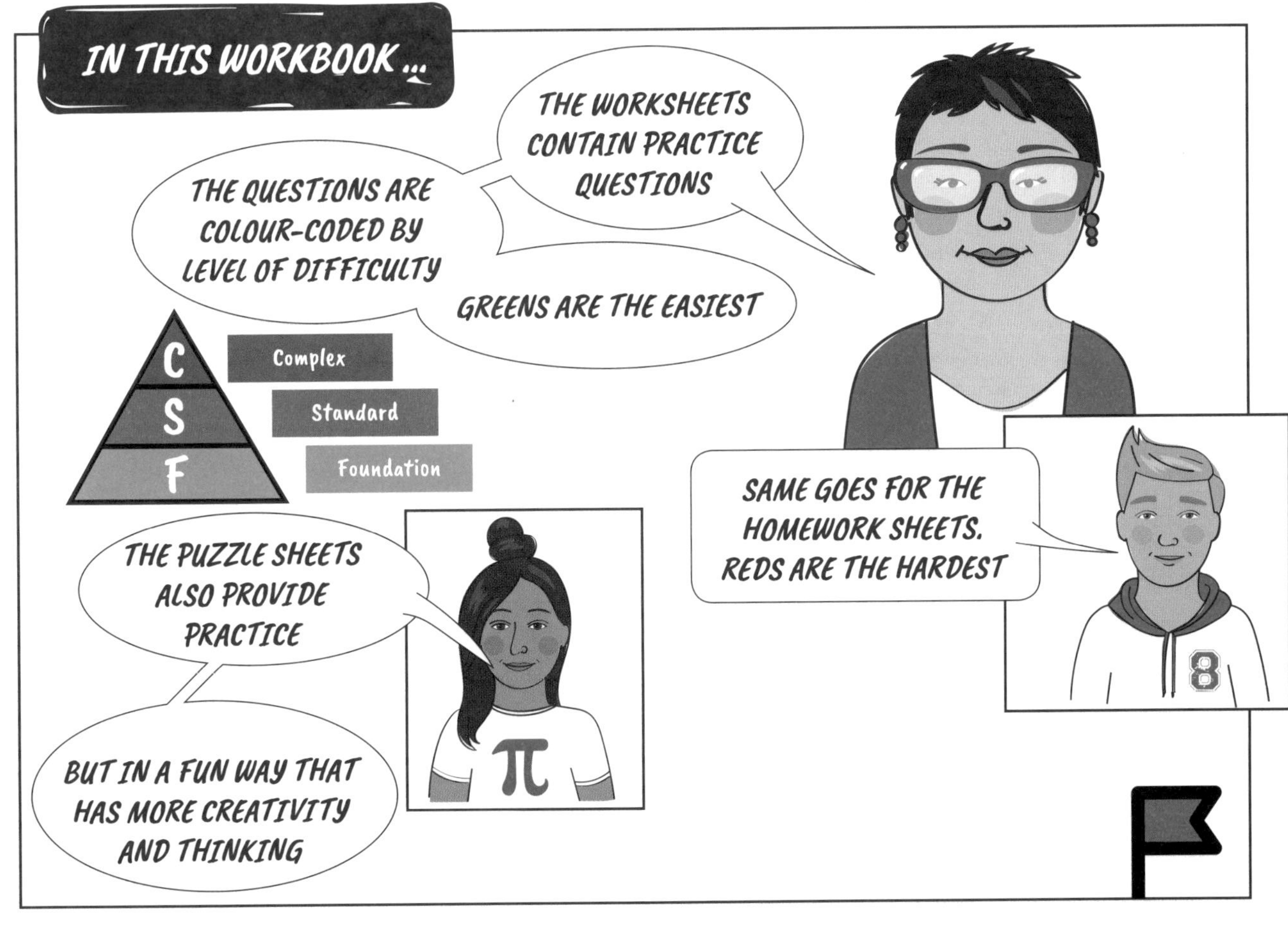

SO LET'S GET GOING.

CURRICULUM GRID

Chapter and content	Australian curriculum strand and substrand
1 INTEREST AND DEPRECIATION	**NUMBER AND ALGEBRA**
Simple interest, Compound interest, Compound interest formula, Depreciation	Money and financial mathematics
2 GRAPHING LINES	**NUMBER AND ALGEBRA**
Length, midpoint and gradient of an interval, Parallel and perpendicular lines, Graphing linear equations, The gradient-intercept equation $y = mx + c$, The general form of a linear equation $ax + by + c = 0$, Finding the equation of a line, Equations of parallel and perpendicular lines	Linear and non-linear relationships
3 SURFACE AREA AND VOLUME	**MEASUREMENT AND GEOMETRY**
Areas of composite shapes, Surface area of a prism, Surface area of a cylinder, Surface areas of composite solids, Volumes of prisms and cylinders	Using units of measurement
4 ALGEBRA	**NUMBER AND ALGEBRA**
The index laws, Adding and subtracting algebraic fractions, Multiplying and dividing algebraic fractions, Expanding and factorising expressions, Expanding binomial products, Factorising quadratic expressions $x^2 + bx + c$	Patterns and algebra
5 COMPARING DATA	**STATISTICS AND PROBABILITY**
The shape of a frequency distribution, Quartiles and interquartile range, Box plots, Parallel box plots, Comparing data sets, Scatterplots, Bivariate data involving time, Statistics in the media	Data representation and interpretation
6 EQUATIONS AND INEQUALITIES	**NUMBER AND ALGEBRA**
Equations, Equations with algebraic fractions, Quadratic equations $x^2 + bx + c = 0$, Equation problems, Equations and formulas, Graphing inequalities on a number line, Solving inequalities	Linear and non-linear relationships
7 GRAPHING CURVES	**NUMBER AND ALGEBRA**
Direct proportion, Inverse proportion, Conversion graphs, The parabola $y = ax^2 + c$, The exponential curve $y = a^x$, The circle $x^2 + y^2 = r^2$, Identifying graphs	Linear and non-linear relationships Real numbers
8 TRIGONOMETRY	**MEASUREMENT AND GEOMETRY**
Pythagoras' theorem, The trigonometric ratios, Finding an unknown side, Finding an unknown angle, Angles of elevation and depression, Bearings, Problems involving bearings	Pythagoras and trigonometry
9 SIMULTANEOUS EQUATIONS	**NUMBER AND ALGEBRA**
Solving simultaneous equations graphically, The elimination method, The substitution method, Problems involving simultaneous equations	Linear and non-linear relationships
10 PROBABILITY	**STATISTICS AND PROBABILITY**
Relative frequency, Venn diagrams, Two-way tables, Tree diagrams, Selecting with and without replacement, Dependent and independent events, Conditional probability	Chance
11 GEOMETRY	**MEASUREMENT AND GEOMETRY**
Congruent triangle proofs, Proving properties of triangles and quadrilaterals, Similar figures, Finding unknown sides in similar figures, Tests for similar figures	Geometric reasoning
12 PRODUCTS AND FACTORS (10A)	**NUMBER AND ALGEBRA**
Perfect squares, Difference of 2 squares, Factorising special binomial products, Factorising quadratic expressions $ax^2 + bx + c$, Factorising algebraic fractions	Patterns and algebra Linear and non-linear relationships
13 SURDS (10A)	**NUMBER AND ALGEBRA**
Surds and irrational numbers, Simplifying surds, Adding and subtracting surds, Multiplying and dividing surds, Binomial products involving surds, Rationalising the denominator	Real numbers
14 QUADRATIC EQUATIONS AND THE PARABOLA (10A)	**NUMBER AND ALGEBRA**
Quadratic equations, Completing the square, the quadratic formula, Quadratic equation problems, the parabola $y = ax^2 + bx + c$, The axis of symmetry and vertex of a parabola	Linear and non-linear relationships
15 FURTHER TRIGONOMETRY (10A)	**MEASUREMENT AND GEOMETRY**
The trigonometric functions, Trigonometric equations, The sine rule, The cosine rule, The area of a triangle $A = \frac{1}{2}ab \sin C$	Pythagoras and trigonometry

HERE ARE SOME MATHS SKILLS YOU NEED FOR THIS YEAR. PART A IS MIXED SKILLS, PART B IS FOR THE FINANCIAL MATHS TOPIC WE'RE LEARNING.

STARTUP ASSIGNMENT 1

PART A: BASIC SKILLS

/ 15 marks

1 Calculate, correct to 3 decimal places:

$$\frac{6 \times 9.2}{\sqrt{5 - 2.35}}$$ ________

2 Simplify:

a $(4x^2)^3$ ________

b $\left(\frac{2}{3}\right)^3$ ________

3 Expand and simplify:

$2(3d - 5) + 4(d + 4)$

____

____

4 What is the formula $V = \pi r^2 h$ used for?

____

5 Solve $3x^2 = 48$.

____

6 Simplify:

a $8 : 28$ ________

b $\frac{4}{5} : \frac{2}{3}$ ________

7 Which congruence test proves that $\triangle ADC \equiv \triangle BCD$ in the rectangle below?

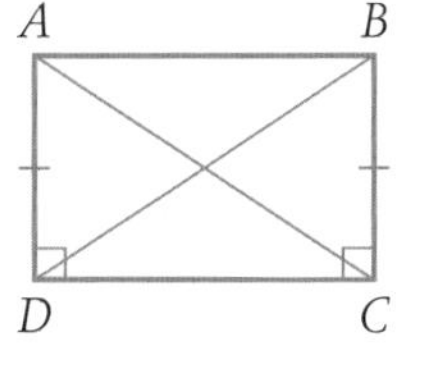

____

8 Convert 1.5 mL/second to L/hour. ________

____

9 Find as a surd the distance between (1, 1) and (3, 5) on the number plane. ________

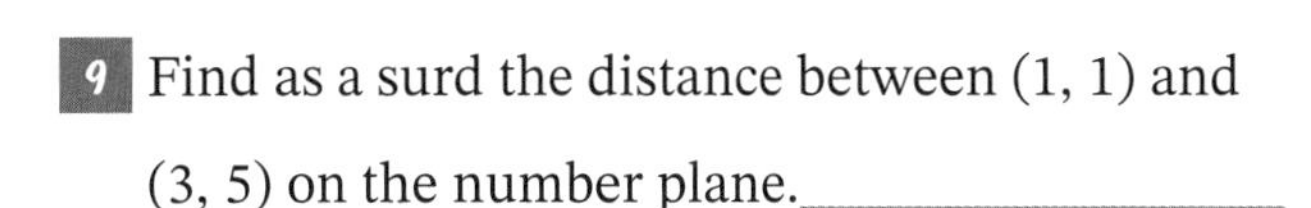

____

10 Find the area of this trapezium.

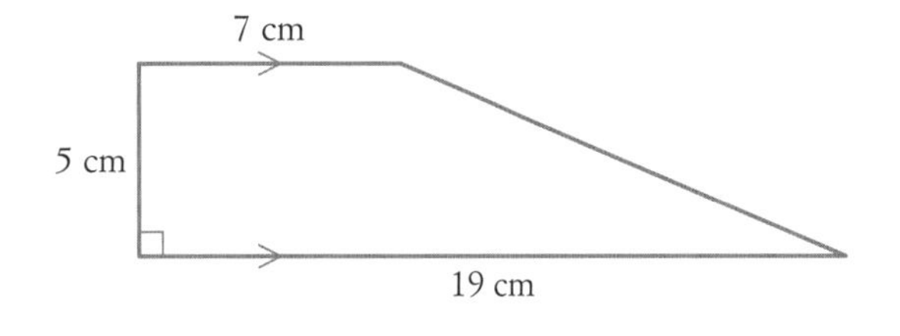

____

____

11 Do the diagonals of a parallelogram bisect each other at right angles?

____

12 Find the value of r in the diagram below.

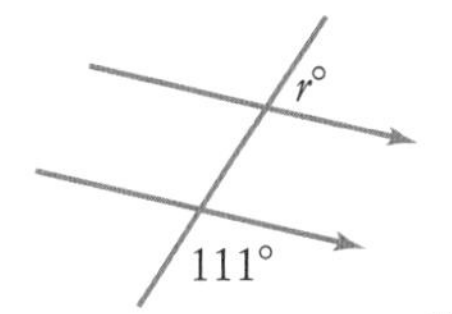

____

13 Evaluate:

$$\frac{4.96 \times 10^7}{3.1 \times 10^{-4}} =$$ ________

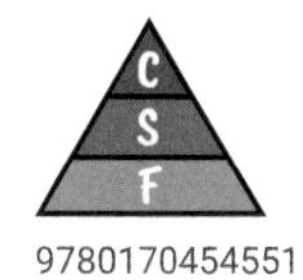

WS WORKSHEET 1

PART B: INTEREST

/ 25 marks

14 Complete:

a $2\frac{1}{2}$ years = ________ months

b $2\frac{1}{2}$ years = ________ weeks

c 1 quarter = ________ months

d 4 years = ________ quarters

e 9% p.a. = ________% per month

f 7% p.a. = ________% per quarter

15 Write each percentage as a fraction.

a 12% ________

b $12\frac{1}{2}$% ________

16 Write each percentage as a decimal.

a 9% ________ **b** $6\frac{1}{2}$% ________

c 1.5% ________ **d** 0.8% ________

17 What percentage is $851 of $3700?

____

18 Evaluate:

a 8% of $4000 ________

b 0.05% of $6420 ________

c $5\frac{3}{4}$% of $30 000 ________

19 To pay off a loan, Chris pays $138 per fortnight. How much does he pay over $1\frac{1}{2}$ years?

____

20 Calculate Sandy's total pay if she works 39 hours at $19.20 per hour and 4 hours overtime at time-and-a-half. ________

____

21 Increase $135 by 7%. ________

____

22 Calculate the simple interest earned if $3000 is invested at 4% p.a. for 5 years.

____

23 Toni needs to repay a loan of $2076 over 2 years. Calculate how much she should pay each month.

____

24 Calculate, correct to the nearest cent:

a $\$950 \times (1.12)^7$ ________

b $\$5300 \times (0.85)^3$ ________

25 Each week, Ivan pays $264.60 in tax. If this is 35% of his income, then calculate his income.

____

____

26 Decrease $2700 by 13%. ________

____

PART C: CHALLENGE

Bonus / 3 marks

When Simon was born, his grandparents deposited $5000 into a trust fund for him. Due to interest, the amount in the fund increases by 6% each year. After how many years will the $5000 double in value?

C
S
F

9780170454551

COMPOUND INTEREST 1

WHEN INTEREST IS COMPOUNDED, YOU GET INTEREST ON YOUR INTEREST.

WORKSHEET

Write all money answers correct to the nearest cent. Use 1 year = 52 weeks = 365 days.

1 How many months in:

a 4 years? ________ **b** 7 years? ________

c $3\frac{1}{2}$ years? ________ **d** $1\frac{1}{4}$ years? ________

2 Write as a decimal:

a 4% ________ **b** 6.5% ________

c 13.25% ________ **d** 8.03% ________

e $7\frac{1}{2}\%$ ________ **f** $3\frac{3}{4}\%$ ________

g 0.6% ________ **h** 0.148% ________

3 Convert each rate to a monthly interest rate expressed as a decimal.

a 13.5% p.a. ________

b 18% p.a. ________

c 3.15% p.a. ________

d 11.07% p.a. ________

4 Convert each rate to a daily interest rate expressed as a decimal correct to 4 significant figures.

a 15% p.a. ________

b 7.5% p.a. ________

c 21.6% p.a. ________

d 16.72% p.a. ________

5 Convert each rate to a quarterly interest rate expressed a decimal.

a 18% p.a. ________

b 7% p.a. ________

c 10.25% p.a. ________

d 15.46% p.a. ________

6

Compound interest formula

$A = P(1 + r)^n$

$5000 is invested at 8% p.a. compounded yearly. Find the final amount at the end of:

a 2 years ________

b 4 years ________

c 5 years ________

C S F

WS WORKSHEET

7 Calculate the final amount of each investment accumulating compound interest.

a $3500 invested at 7% p.a. for 5 years

____

b $9900 invested at 10.2% p.a. for 7 years

____

c $25 000 invested at 8.95% for 3 years

____

d $12 000 invested at 4% p.a. compounded half-yearly for 2 years

____

e $15 500 invested at 9.6% p.a. compounded monthly for 4 years

____

f $37 000 invested at 11.7% p.a. compounded quarterly for 3 years

____

8 Calculate the compound interest earned when:

a $4000 is invested at 6% p.a. for 4 years

____

b $21 500 is invested at 8.5% p.a. compounded half-yearly for 6 years

____

c $32 000 is invested at 9.2% p.a. compounded quarterly for 2 years

____

d $9750 is invested at 7.05% p.a. compounded monthly for $1\frac{1}{2}$ years

____

9 Calculate the principal to be invested at compound interest to reach each final amount.

a $3200 in 3 years invested at 9% p.a.

____

b $9500 in 4 years invested at 7.3% p.a.

____

c $15 000 in 2 years invested at 8% p.a.

____

10 Find the values missing from this table.

	Principal	Rate (% p.a.)	Time	Compounded	Final amount	Interest
a	$5500	7%	4 years	Yearly		
b		6.4%	6 years	Half-yearly	$9300	
c	$20 000	12.6%	3.5 years	Monthly		
d		9%	2 years	Monthly	$25 600	
e		14.8%	$2\frac{3}{4}$ years	Quarterly	$12 220	

C S F

9780170454551

DEPRECIATION 1

DEPRECIATION IS THE OPPOSITE OF COMPOUND INTEREST, WHEN THE VALUE OF AN ITEM DECREASES BY THE SAME PERCENTAGE.

WS WORKSHEET

1 The purchase price of a boat is \$48 000.
If the boat depreciates by 10% p.a., calculate its value after 5 years.

______

2 A \$34 000 new car depreciates by 11% p.a.

a Calculate the value of the car after 8 years.

______

b Calculate the amount by which the vehicle depreciates in 8 years.

______

3 A local business has purchased office furniture to the value of \$26 500.
Find the furniture's value after 3 years of use if it depreciates at 28.5% per annum.

______

4 A factory depreciates in value by 4.5% per annum. If its current value is \$315 000, find its value after 10 years.

______

5 A construction company purchases a new grader for \$356 400. If the grader depreciates at a rate of 22% p.a.:

a calculate its value after 1 year

______

b calculate its value after the second year

______

c by how much has the grader depreciated in the second year?

______

6 Thomas buys a new snowmobile for \$11 820. It depreciates at a rate of 15.5% per annum.

a Find the value of Thomas' snowmobile after 2 years.

______

b Find the value of Thomas' snowmobile after 5 years as a percentage of its original value, correct to one decimal place.

______

7 A Blu-ray DVD player, originally costing \$480, depreciates at 10% per annum. What percentage of its original value is its value after:

a 1 year? ______

b 2 years? ______

c 5 years? ______

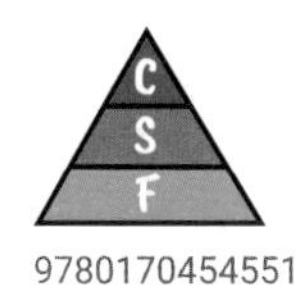

WS WORKSHEET

8 Patrick spent $18 500 on equipment for his gardening business. The equipment depreciates at 18.5% per year.

a Find the value of the equipment after 4 years.

____

b Find the amount of depreciation of the equipment after 2 years.

____

c Find how long it will take Patrick's equipment to have a value less than $5000.

____

9 A company car purchased for $46 000 depreciates at 11.4% per annum.

a Calculate the value of the car after 3 years.

____

b Calculate the total depreciation over the first 7 years.

____

c How long will it take for the car to reach a value below $10 000?

____

____

10 A front-end loader was purchased for $415 300. It depreciates at 20% per annum. The owner sells the loader for $245 000 after 3 years. Was there a profit or loss made on the sale of the front-end loader?

____

____

____

____

11 The Australian Taxation Office allows depreciation on tools of trade as a legitimate tax deduction. The depreciation rate is 13.5% per annum. A bricklayer purchases tools to the value of $18 300. When the value falls below $6000, the bricklayer is allowed to write off the tools on the next year's tax return. Calculate when the tools can be written off.

____

____

____

____

12 Adrian bought a new car for $27 900. The salesperson claimed that, at 10% p.a. depreciation, the car would decrease in value by $2790 per year. Do you agree or disagree with this statement? Give reasons.

____

____

____

____

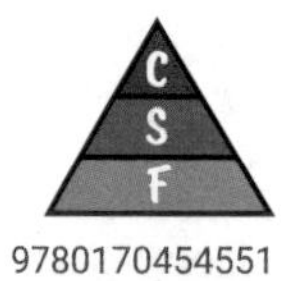

9780170454551

INTEREST AND DEPRECIATION CROSSWORD 1

UNSCRAMBLE THE WORDS FROM THIS TOPIC NEXT PAGE TO COMPLETE THIS PUZZLE.

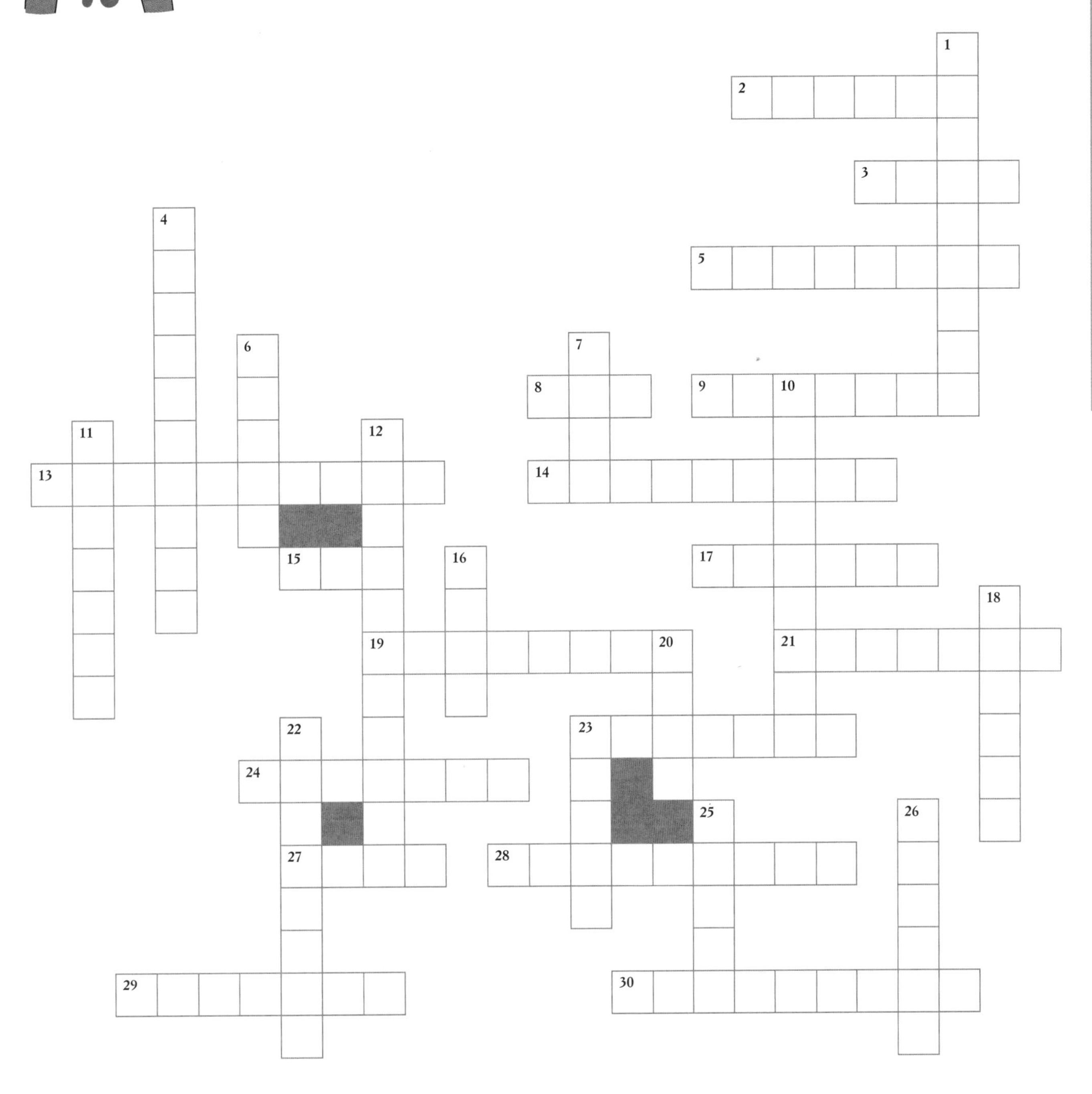

PUZZLE SHEET

PS PUZZLE SHEET

Clues across

1 PETMARYEN
4 TENSTENVIM
6 SORGS
7 AGEW
10 NIPLIPCAR
11 DAGLION
12 TRYGOLFHINT
16 TEAR
18 LAANNU
20 METR
22 CUPNODOM
23 LAFIN
25 VEALE
26 ALYSRA

Clues down

2 IMPELS
3 TALF
5 ROVEITEM
8 ATX
9 TOPSIDE
13 MINICOSMOS
14 ICEDDONUT
15 ENT
17 NOMICE
19 NITTREES
21 TNCREEP
23 LAFORUM
24 MYLONTH
27 GAPY
28 LUERYTQAR
29 LACANBE
30 WOKRECIPE

Name:

Due date:

Parent's signature:

Part	Marks
Part A	/ 8 marks
Part B	/ 8 marks
Part C	/ 8 marks
Part D	/ 8 marks
Total	/ 32 marks

COMPOUND INTEREST 1

THIS IS YOUR WEEKLY HOMEWORK ASSIGNMENT, COVERING THE CURRENT TOPIC AS WELL AS MIXED REVISION. NO CALCULATORS IN PART A!

PART A: MENTAL MATHS

Calculators not allowed

1 Find 5% of \$4000. ______

2 Between which 2 consecutive whole numbers is the value of $\sqrt{70}$?

______

3 Calculate, as a surd, the distance between the points (−3, −2) and (−2, 7) on the number plane.

______

4 Write the formula for a closed cylinder's:

a surface area ______

b volume ______

5 For the data in the table, find:

x	y
2	3
3	8
4	6
5	6
Total	23

a the median ______

b the mode ______

c the range ______

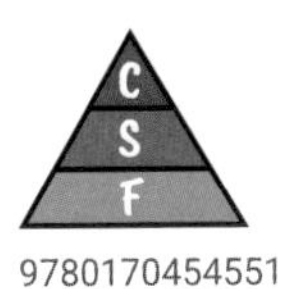

PART B: REVIEW

1 Convert each percentage to a decimal.

a 17.5% ______

b 5% ______

2 Increase \$4500 by 2.5% ______

3 Complete:

a 1 year = ______ weeks

b 1 year = ______ months

c 1 year = ______ days

4 Evaluate correct to the nearest cent:

a $\$50\,000 \times (1.05)^5$ ______

b $\$34\,300 \times (1.071)^{10}$ ______

HW HOMEWORK

PART C: PRACTICE

› Simple interest
› Compound interest

1 Calculate the simple interest earned on $25 000 invested for 5 months at 5.3% p.a.

2 (2 marks) The simple interest on a loan of $15 960 over 5 years is $5200.

Calculate the interest rate p.a., correct to one decimal place.

3 Calculate the value of an investment of $18 250 at 3.2% p.a. interest compounded annually after 3 years.

4 For an investment of $20 000 for 4 years compounded monthly at 4.5% p.a., find:

a the total amount

b the interest earned

5 (2 marks) $7400 is invested for 2 years at 3.8% p.a., compounded quarterly. Calculate the interest earned.

PART D: NUMERACY AND LITERACY

1 a What type of interest is calculated only on the original principal?

b What type of interest is used in the formula $A = P(1 + r)^n$?

c In the formula $I = Prn$, what does the n stand for?

d In the formula $A = P(1 + r)^n$, what does the P stand for?

2 $38 000 is invested for 6 months at 17.5% p.a.

a Calculate the simple interest.

b Calculate the total amount.

3 For an investment of $16 500 for one year at 1.6% p.a. compounded daily, calculate:

a the total amount

b the compound interest

HW HOMEWORK

9780170454551

STARTUP ASSIGNMENT 2 (2)

HI, I'M MS LEE. THIS ASSIGNMENT REVISES YOUR NUMBER PLANE AND GRAPHING SKILLS.

WS WORKSHEET

PART A: BASIC SKILLS

/ 15 marks

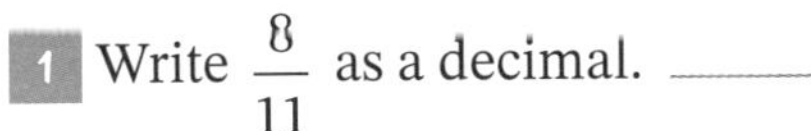

1 Write $\frac{8}{11}$ as a decimal. ______________

2 Find the simple interest earned on $5000 invested for $3\frac{1}{2}$ years at 6.2% p.a.

____

3 Complete: 1 m^3 can hold ________ L.

4 Is $\sqrt{2.5}$ rational or irrational?

____

5 Evaluate $\frac{9}{2\sqrt{3}}$, correct to 3 significant figures.

____

6 How much petrol can I buy for $50 at 142.3 cents/litre? Answer to the nearest 0.1 L.

____

7 Write the formula for the surface area of a cylinder.

____

8 **a** Write 0.041 05 in scientific notation.

____

b How many significant figures has 0.041 05?

____

9 Expand $3(3y + 1)(3y - 1)$. ______________

____

10 What is the angle sum of a quadrilateral?

____

____

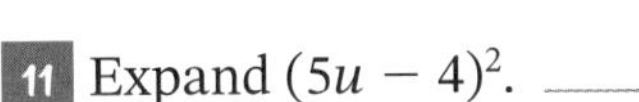

11 Expand $(5u - 4)^2$. ______________

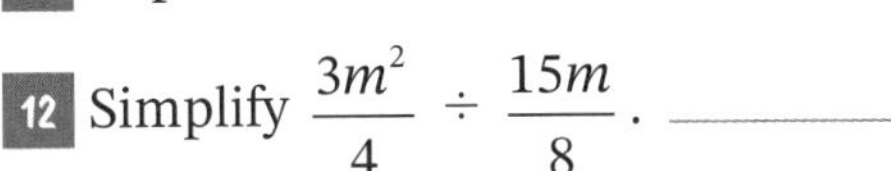

12 Simplify $\frac{3m^2}{4} \div \frac{15m}{8}$. ______________

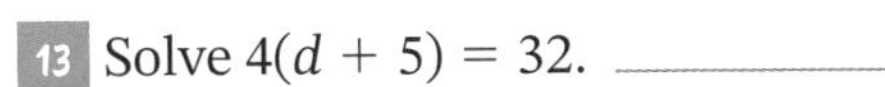

13 Solve $4(d + 5) = 32$. ______________

____

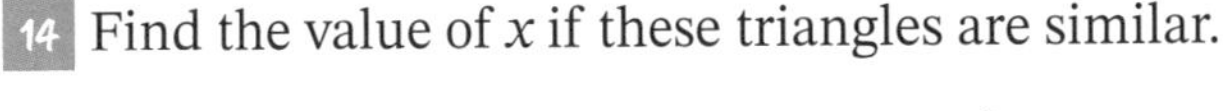

14 Find the value of x if these triangles are similar.

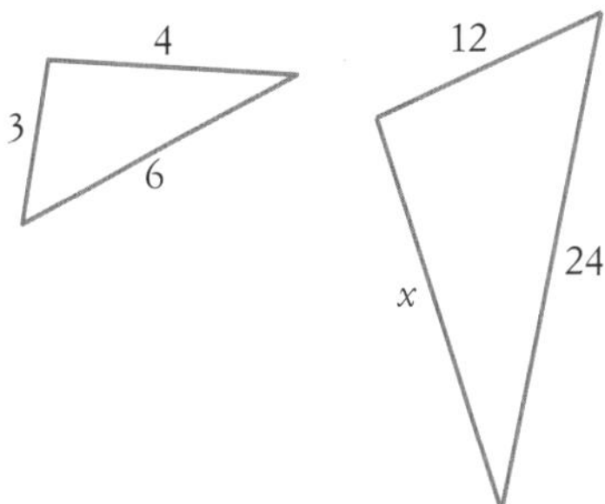

PART B: COORDINATE GEOMETRY

/ 25 marks

15 Complete the table below for $y = 4x - 5$.

x	−2	−1	0	1
y				

16 Evaluate $\frac{d - b}{c - a}$ if $a = -7$, $b = 2$, $c = -1$ and $d = 0$.

____

9780170454551

WS WORKSHEET

Questions **17** to **19** refer to this diagram.

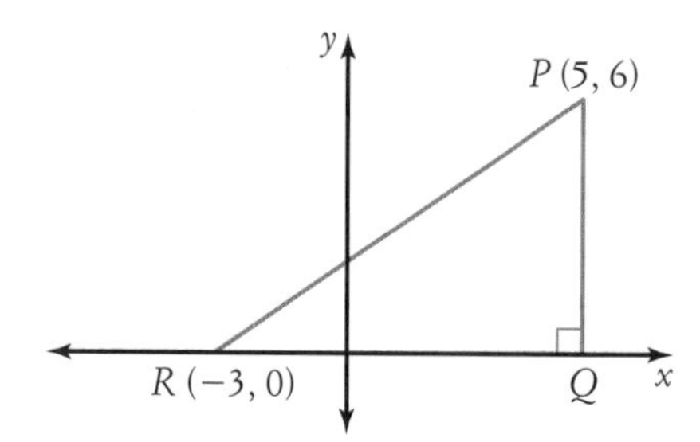

17 If $\triangle PQR$ is right-angled, write the coordinates of Q. ______

18 Find the area of $\triangle PQR$. ______

19 For the hypotenuse RP, find its:

a length ______

______

b midpoint ______

______

c gradient. ______

______

20 Graph $y = -2$ and $y = -3x$ on a number plane.

21 Write one property of the sides of a:

a trapezium

______

______

b rhombus

______

______

c parallelogram.

______

______

22 Find the length of the hypotenuse of a right-angled triangle as a surd if the other sides are 3 cm and 6 cm. ______

______

23 Write the equation of the horizontal line that goes through $(0, -4)$.

______

24 Draw a line whose gradient is negative.

C S F

9780170454551

25 Graph $y = 2x - 3$ on a number plane and find its gradient and y-intercept.

____

26 Find y when $x = 2$ if:

a $y = 8 - 3x$ ______________

b $y = \frac{x}{4} + 5$ ______________

c $y = \frac{1}{2}x^2$ ______________

d $y = x^3 - 4$ ______________

27 Does the point (4, 7) lie on the line $y = -2x + 15$? ______________

____

28 Show that $y = \frac{1}{2}x + 3$ can be rewritten as $2x - 4y + 12 = 0$.

____

____

____

____

____

PART C: CHALLENGE

Bonus / 3 marks

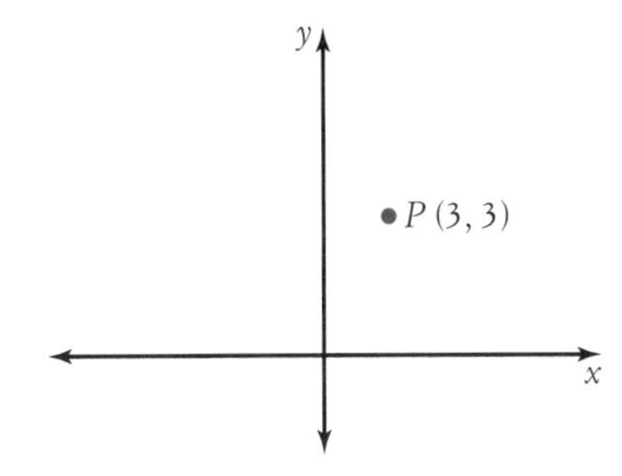

- How many points on the number plane (with integer coordinates) are exactly 5 units from $P(3, 3)$?

____

- If *all* points that are 5 units from P are graphed (including those with non-integer coordinates), what shape would result?

____

C
S
F

9780170454551

② LINEAR EQUATIONS CODE PUZZLE

6 12 7 10   14 15 14   2 17 11   25 7 9 7 8 8 11 8

8 15 17 11   4 7 1   10 2   7 17 2 10 12 11 9

7 10   10 12 11   25 7 9 10 1 ?

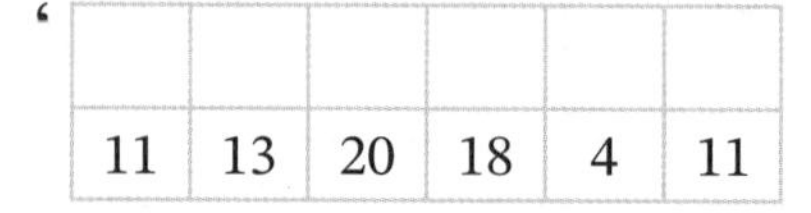

‘11 13 20 18 4 11   21 11,   1 2 18   8 2 2 24

23 11 9 1   5 7 21 15 8 15 7 9   3 18 10   6 11,   23 11

17 11 23 11 9   21 11 10,   12 7 23 11   6 11 ?’

The numbers in the grid above match the question numbers. Write each of the following linear equations in the form $y = mx + c$ (where appropriate) and match each one with its correct feature or graph on the next page. Each answer has a corresponding letter. Fill in the grid above with the letters that match the questions, to decode a riddle.

1 $2x + y - 1 = 0$ ______

2 $x + y + 1 = 0$ ______

3 $x + 2y - 1 = 0$ ______

4 $y = 7$ ______

5 $x = 1$ ______

6 $y = x$ ______

7 $x + 5y = 0$ ______

8 $y = x - 3$ ______

9 $4x - 2y + 6 = 0$ ______

10 $y = \frac{1}{2}x + 1$ ______

11 $y = -2$ ______

12 $2x - 3y + 3 = 0$ ______

13 $2x - y - 2 = 0$ ______

14 $y = -3x$ ______

15 $x = -2$ ______

16 $y = 3x - 4$ ______

17 $x - y + 4 = 0$ ________

18 $8x + 2y - 20 = 0$ ________

19 $2x - 3y - 9 = 0$ ________

20 $2x - y - 1 = 0$ ________

21 $x + y - 2 = 0$ ________

22 $y = x + \frac{1}{2}$ ________

23 $3x - 2y - 14 = 0$ ________

24 $5x + y - 2 = 0$ ________

25 $y = 0$ ________

T

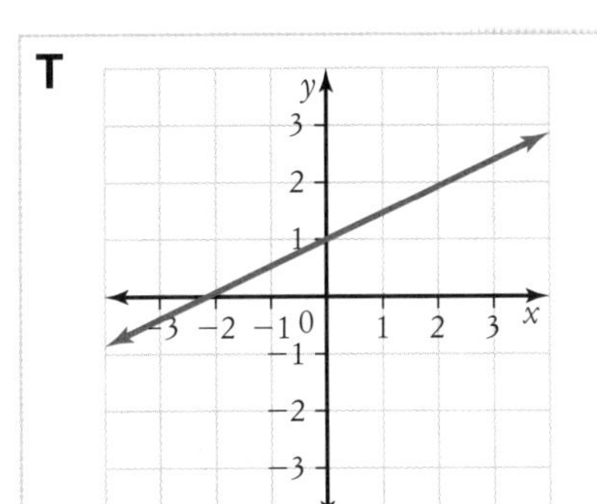

Y

L

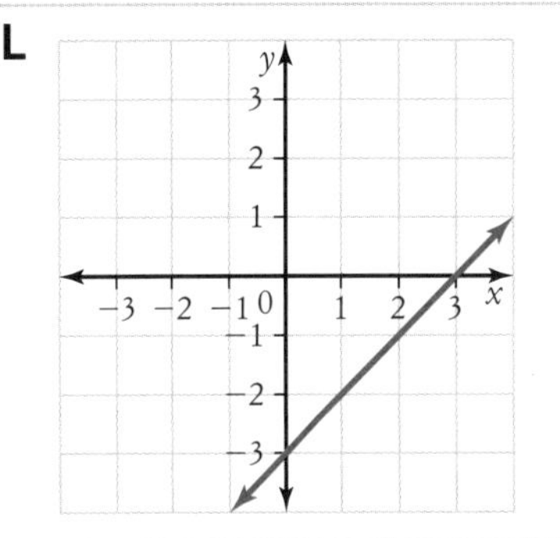

F

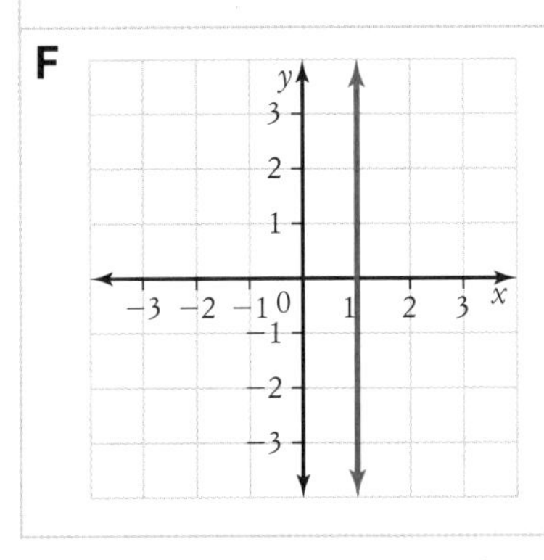

X

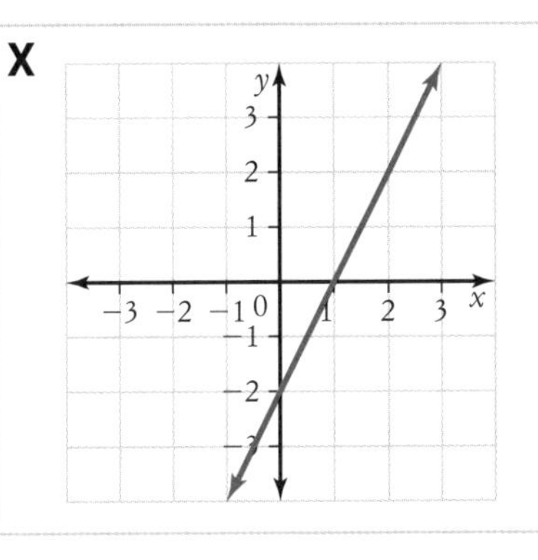

W

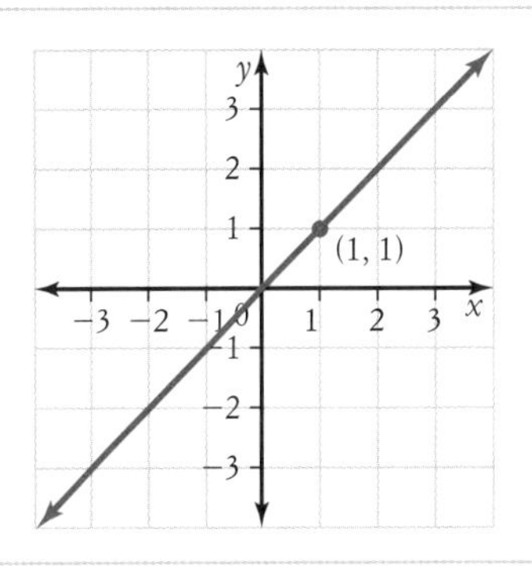

C $m = 2, c = -1$	**K** $m = -5$	**G** $m = 1, c = \frac{1}{2}$	**D** $m = -3$
A $m = -\frac{1}{5}$	**H** $m = \frac{2}{3}, c = 1$	**V** $m = \frac{3}{2}$	**O** $m = c = -1$
N $c = 4$	**S** $c = 7$	**J** $c = -4$	**U** $c = 10$
I vertical line passing through $(-2, 0)$		**M** $y = 2 - x$	**R** $y = 2x + 3$
B $y = -\frac{1}{2}x + \frac{1}{2}$	**Z** $y = \frac{2}{3}x - 3$	**E** $y = -2$	**P** the x-axis

PS PUZZLE SHEET

WS WORKSHEET

2 WRITING EQUATIONS OF LINES

FIND THE EQUATION OF ANY LINE THAT FITS EACH DESCRIPTION, IN THE FORM $y = mx + c$.

1 (1, 3) lies on the line.	**2** Parallel to $y = 6x - 7$	**3** Parallel to the x-axis.	**4** Has a negative gradient and passes through (0, 0).
5 Perpendicular to $y = -3x - 1$.	**6** The y-intercept is 5.	**7** The gradient is $\frac{1}{6}$.	**8** Perpendicular to $3x - 2y + 4 = 0$.
9 Passes through $(-1, 4)$ and is parallel to the y-axis.	**10** Steeper than $y = 2x - 4$.	**11** The x-intercept is -7.	**12** Passes through $\left(\frac{1}{2}, 5\right)$.
13 Parallel to $2x - y + 7 = 0$.	**14** Perpendicular to the x-axis.	**15** Has the same y-intercept as $x - 3y + 15 = 0$.	**16** Perpendicular to the y-axis and passes through $(-3, -5)$.
17 Has a positive gradient and passes through $(0, -5)$.	**18** Parallel to $y = -5x + 7$ and cuts the x-axis at 2.	**19** The x-intercept is 10 and the y-intercept is -3.	**20** Steeper than $y = -\frac{x}{2} - 4$.
21 $(2, -1)$ lies on the line and it has a gradient of 3.	**22** Perpendicular to $y = \frac{2x}{3} + 7$ with a negative y-intercept.	**23** Passes through $(0, -4)$ and $(9, 0)$.	**24** Passes through $(1, -5)$ and $(5, -9)$.

9780170454551

GRAPHING LINES CROSSWORD ②

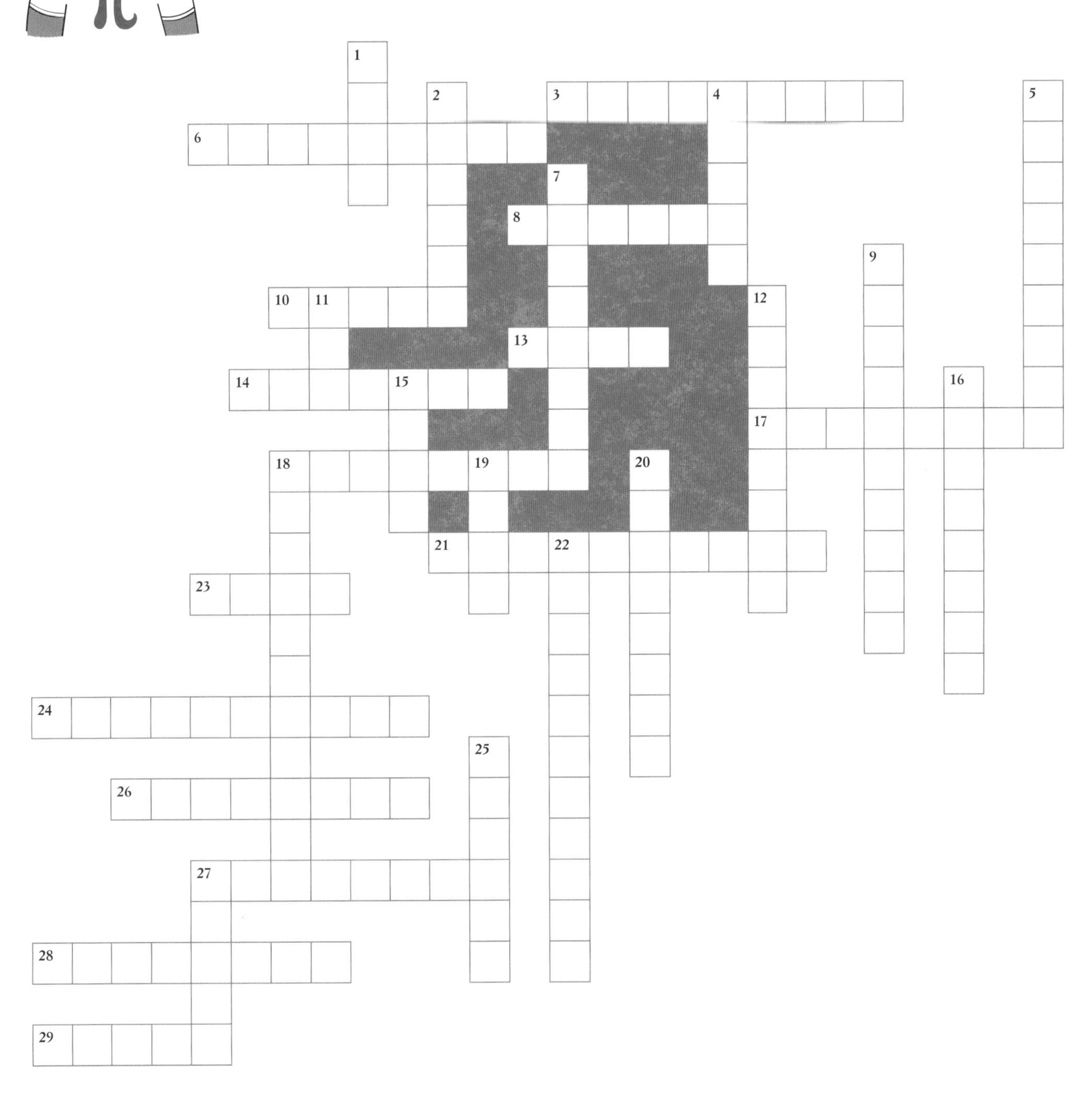

AXES
CARTESIAN
CONSTANT
COORDINATES
EQUATION
EXACT
FORM
GENERAL
GRADIENT
GRAPH
HYPOTENUSE
INCLINATION
INTERCEPT
INTERVAL
LENGTH
LINE
LINEAR
MIDPOINT
NEGATIVE
ORIGIN
PARALLEL
PLANE
POINT
POSITIVE
PYTHAGORAS
RECIPROCAL
RISE
RUN
STEEPNESS
SURD
VERTICAL

2 COORDINATE GEOMETRY

AS IN SPORT, DRILL AND PRACTICE ARE IMPORTANT IN MATHS. PART B OF THIS HOMEWORK ASSIGNMENT IS REVIEW OF PREVIOUS WORK, PART C IS PRACTICE OF THIS WEEK'S WORK.

Name:

Due date:

Parent's signature:

Part A	/ 8 marks
Part B	/ 8 marks
Part C	/ 8 marks
Part D	/ 8 marks
Total	/ 32 marks

PART A: MENTAL MATHS

Calculators not allowed

1 A TV costs \$400 after 20% discount. What was its original price?

2 A fridge has its price reduced from \$800 to \$680. What is the percentage discount?

3 Convert:

a 60 km/h = ______ m/min

b 4 mL/min = ______ L/h

4 (4 marks) Find the value of each variable, giving reasons.

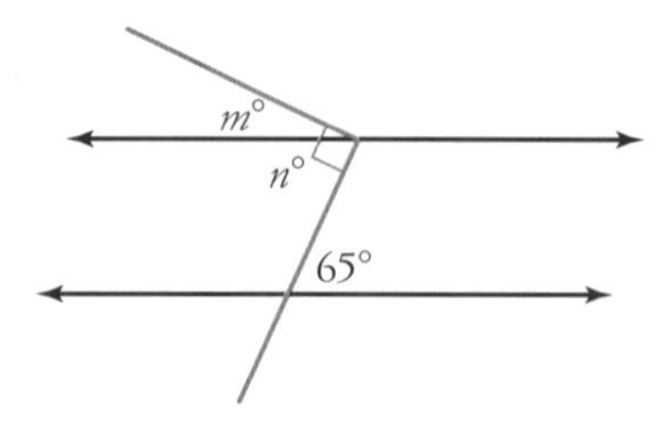

PART B: REVIEW

1 (4 marks) Complete each table of values.

a $y = x - 2$

x	−2	−1	0	1
y				

b $y = 4x - 3$

x	−1	0	1	2
y				

2 State whether each line's gradient is positive, negative or neither.

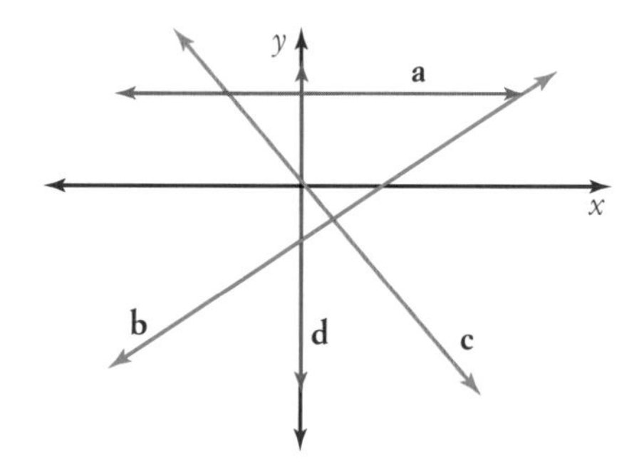

a ______

b ______

c ______

d ______

C S F

9780170454551

PART C: PRACTICE

› Length, midpoint, gradient
› Graphing $y = mx + c$

1

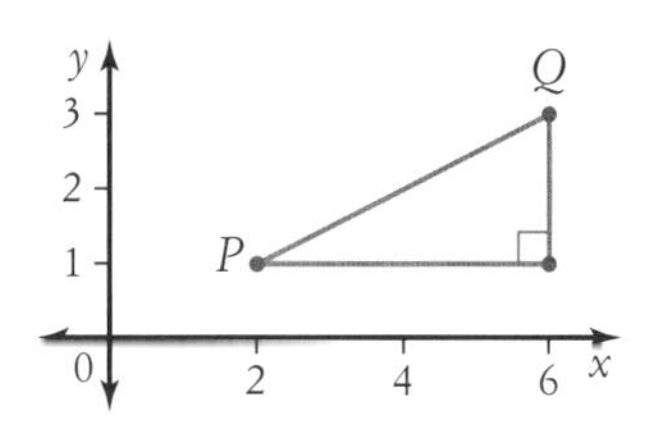

a Find the length of PQ as a surd.

____

____

b Find the midpoint of PQ.

____

c Find the gradient of PQ.

____

2 (3 marks) Graph $y = -3x + 6$ on the number plane and write its x-intercept and y-intercept.

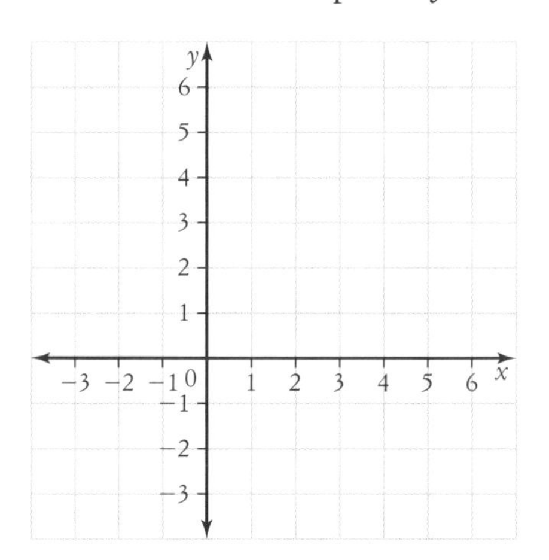

____

____

____

____

3 What is the gradient of the line with equation $y = -3x + 6$?

____

4 Test whether $(5, -9)$ lies on the line with equation $y = -3x + 6$.

____

PART D: NUMERACY AND LITERACY

1 **a** Draw a quadrilateral $ABCD$ with vertices at $A(0, 4)$, $B(6, 1)$, $C(2, -3)$, $D(-2, -1)$.

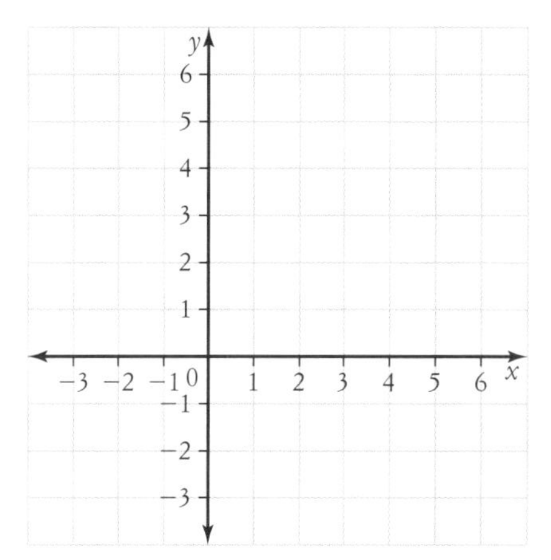

b Find the gradient of the sides of $ABCD$.

$m_{AB} =$ ______________________________

$m_{CD} =$ ______________________________

$m_{AD} =$ ______________________________

$m_{BC} =$ ______________________________

c What type of quadrilateral is $ABCD$?

____

2 Write the equation of a line with gradient -1 and y-intercept -2.

____

3 Write the equation of the vertical line that goes through $(3, -4)$.

____

HW HOMEWORK

C S F

2 GRAPHING LINES

Name:

Due date:

Parent's signature:

PART D QUESTIONS CAN BE COMPLEX BECAUSE THEY ASK YOU TO WRITE ABOUT YOUR MATHS, USING THE RIGHT TERMINOLOGY.

Part A	/ 8 marks
Part B	/ 8 marks
Part C	/ 8 marks
Part D	/ 8 marks
Total	/ 32 marks

PART A: MENTAL MATHS

Calculators not allowed

1 Factorise each expression.

a $49m^2 - 35m$

b $8(x + 2) - p(2 + x)$

2 Simplify each expression.

a $\dfrac{56x^8y^6}{(2x^2y^2)^2}$

b $\left(\dfrac{3}{4m^2}\right)^{-2}$

3 For this stem-and-leaf plot, identify any:

Stem	Leaf
1	2 5 6
2	2 3 6 7 8
3	0 4 4 8 9 9 9
4	1 2 4 5 5 6 7 8 8 9
5	0 1
6	9

a outliers

b clusters

4 (2 marks) If $P = 50$ and $L = 4$, find W if $P = 2(L + W)$.

PART B: REVIEW

1 **a** Find the gradient of the line with equation $y = 2x - 6$.

b What is the gradient of a line that is perpendicular to $y = 2x - 6$?

2 Write the equation of a line with gradient $\frac{1}{2}$ and y-intercept 7.

C S F

HW HOMEWORK

3 Graph the line $y = -2$ on the number plane.

4 (4 marks) Find the gradients of each pair of lines, and then determine whether they are parallel or perpendicular, giving reasons.

a $y = 4x + 13$ $\quad m_1 =$ ________

$y = 4x - 3$ $\quad m_2 =$ ________

____

b $2x + 4y - 16 = 0$ $\quad m_1 =$ ________

$-6x + 3y + 3 = 0$ $\quad m_2 =$ ________

____

____

____

PART C: PRACTICE

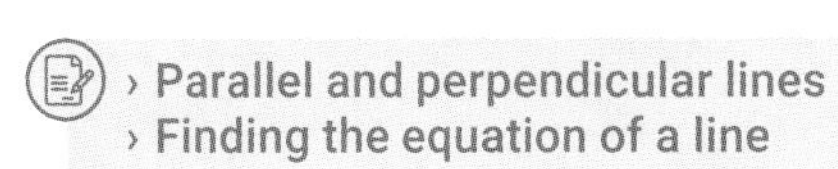

1 Find the equation of this line.

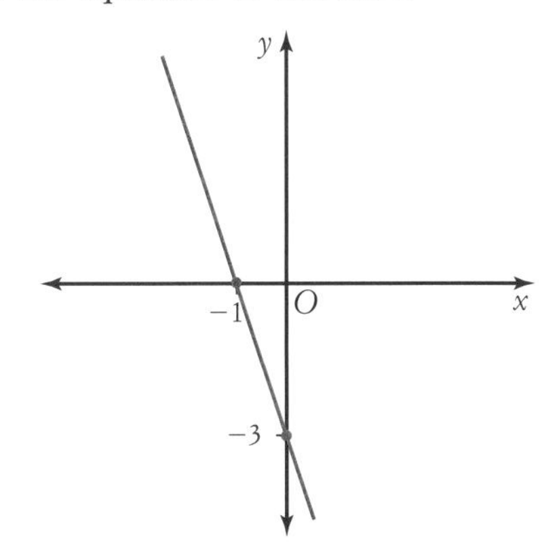

2 (3 marks) Find the equation of the line that is parallel to:

a $y = 5x - 2$ and has a y-intercept of 3

____

____

b $2x - 2y - 8 = 0$ and passes through $(2, 1)$

____

____

3 (4 marks) Find the equation of the line that is perpendicular to:

a $y = 5x - 2$ and has a y-intercept of 3.

____

____

b $y = -3x - 4$ and passes through the midpoint of $(6, -8)$ and $(0, -4)$

____

____

PART D: NUMERACY AND LITERACY

1 Complete: If 2 lines have gradients m_1 and m_2 and

a $m_1 = m_2$, then they are

____

b $m_2 = -\frac{1}{m_1}$, then they are

____

HW HOMEWORK

C S F

2

a Find the gradient of interval AB.

b Find the midpoint of AB.

c The red line is perpendicular to AB and passes through its midpoint. What is its gradient?

d Find the equation of the red line.

3

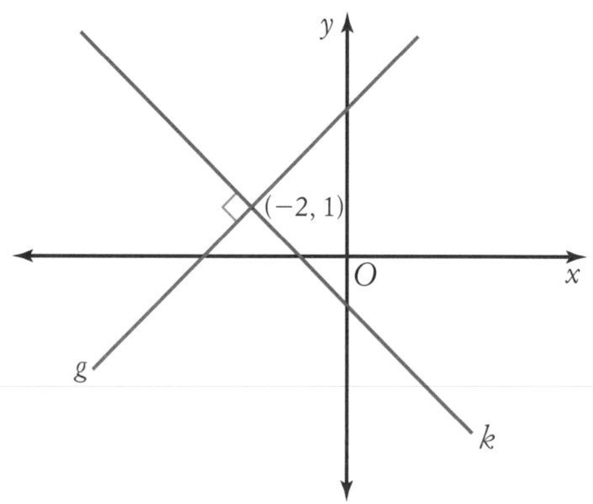

a Find the equation of line g if its gradient is $\frac{1}{2}$.

b Find the equation of line k.

C S F

9780170454551

STARTUP ASSIGNMENT 3 (3)

THIS ASSIGNMENT CAN BE DONE AT THE START OF THE TOPIC BECAUSE IT REVISES AREA AND VOLUME SKILLS YOU'LL NEED FOR THAT TOPIC.

WORKSHEET

PART A: BASIC SKILLS

/ 15 marks

1 Draw an obtuse-angled isosceles triangle.

2 Simplify $(5k)^{-2}$. ______

3 For the values 19, 12, 11, 17, 11, 12, 13, 11, find:

a the mean ______

b the median ______

4 What is the highest common factor of $18a^2b$ and $12abc$? ______

5 Find the size of the angle x in the diagram below. ______

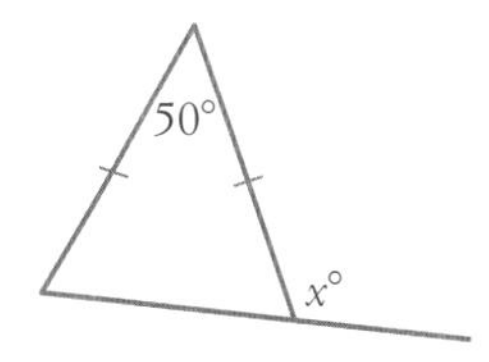

6 Write $\frac{17}{20}$ as a percentage. ______

7 Write 37 800 000 in scientific notation.

______

8 Solve $3x - 5 = x + 9$. ______

9 For the points $A(1, 7)$ and $B(-3, 15)$, find:

a the length of AB, to 2 decimal places

______

b the gradient of AB. ______

10 Simplify $\frac{10ab}{2a^2}$. ______

11 Convert $\frac{5}{6}$ to a decimal. ______

12 Jane earned \$686.70 for selling \$9810 worth of cosmetics. What was her percentage commission? ______

13 Find d if these triangles are similar.

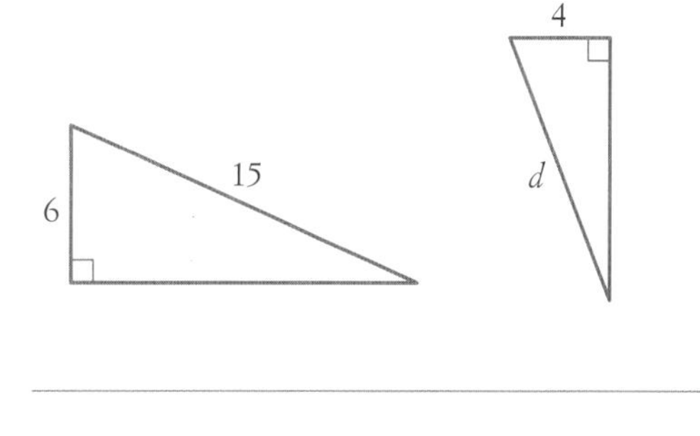

______

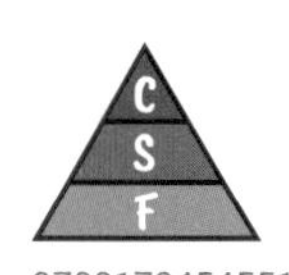

PART B: AREA AND VOLUME

/ 25 marks

Round answers to 2 decimal places where necessary.

14 **a** Draw a square pyramid.

b Draw its net.

15 A circle has a radius of 6 cm. Find, correct to 2 decimal places, its:

a circumference ______

b area ______

16 A cube has sides of length 8 cm. Find its:

a surface area ______

b volume ______

17 **a** How many cm^2 in 1 m^2? ______

b How many litres in 1 m^3? ______

18 **a** Name this solid. ______

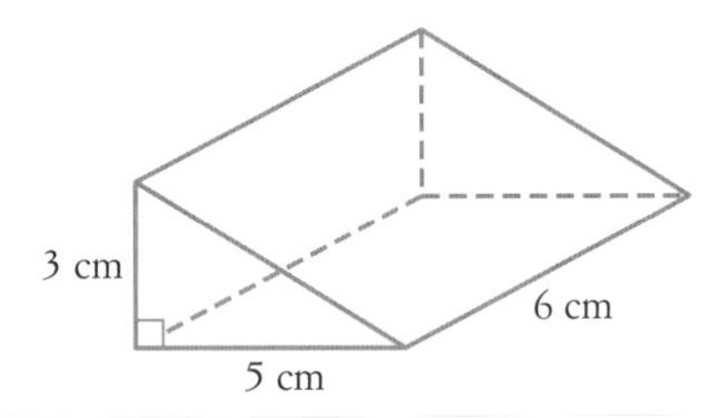

b How many faces does it have? ______

c What shapes are its faces?

______

d Find the volume of the solid. ______

19 If $r = 3$ and $h = 9$, evaluate correct to 2 decimal places:

a $\frac{1}{3}\pi r^2 h$ ______

b $\frac{4}{3}\pi r^3$ ______

20 Write the formula for the area of a trapezium.

______

21 **a** Name this shape.

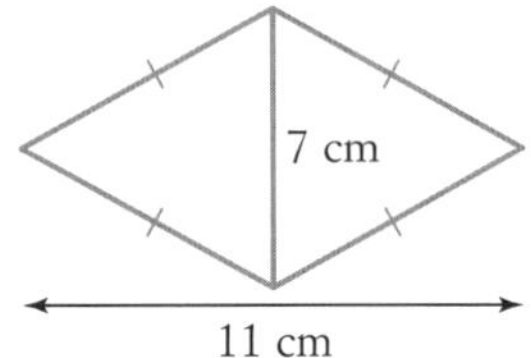

______

b Find the shape's area. ______

22 Find the perimeter of this triangle.

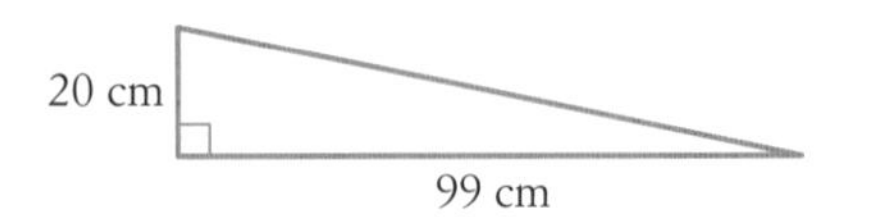

______

23 Find the base length of a parallelogram with a perpendicular height of 4 cm if its area is 28 cm^2. ______

9780170454551

24 Write one difference between a prism and a pyramid.

25 Find the radius of this cone.

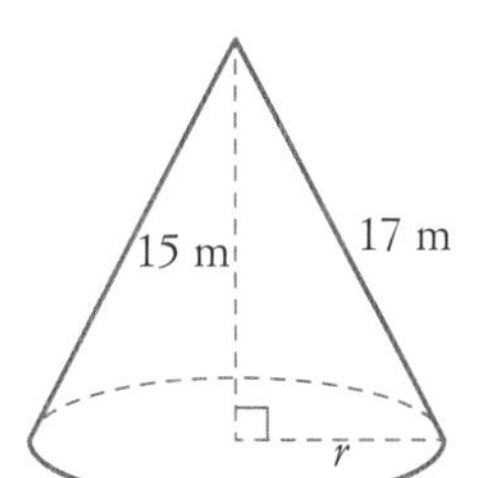

26 Calculate each area below.

a

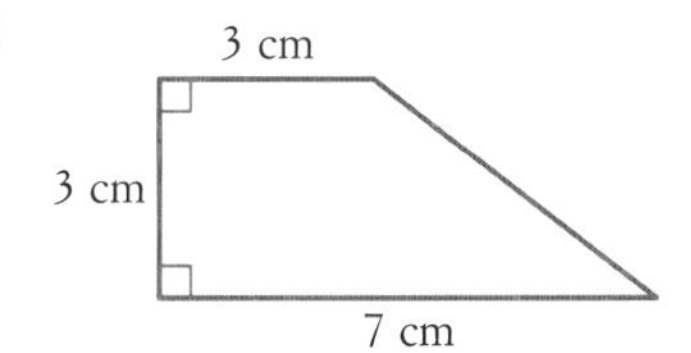

b

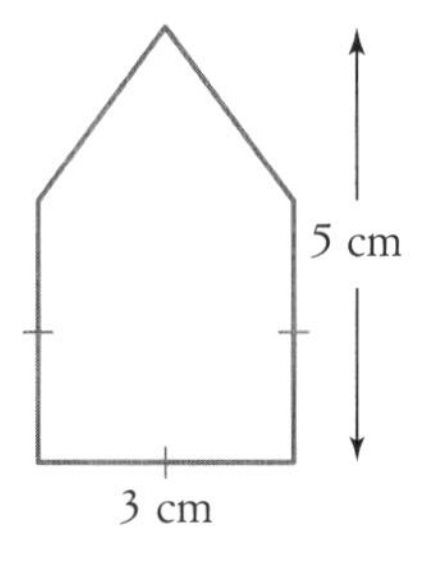

c

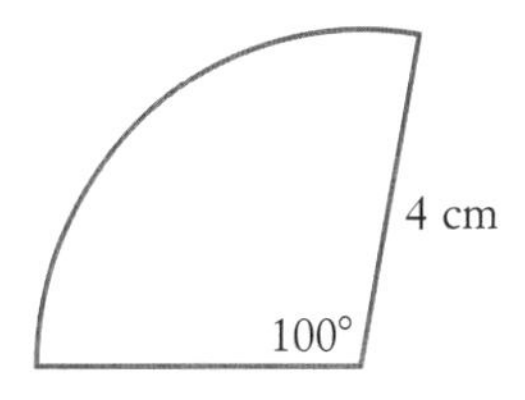

27 Find the length of a rectangle if its length is double its width and its area is 18 cm^2.

PART C: CHALLENGE

Bonus / 3 marks

A cylinder holds 3 tennis balls neatly.

What fraction of the cylinder's volume is taken up by the tennis balls if the volume of a sphere with radius r is $V = \frac{4}{3}\pi r^3$?

3 BACK-TO-FRONT PROBLEMS

WS WORKSHEET

1

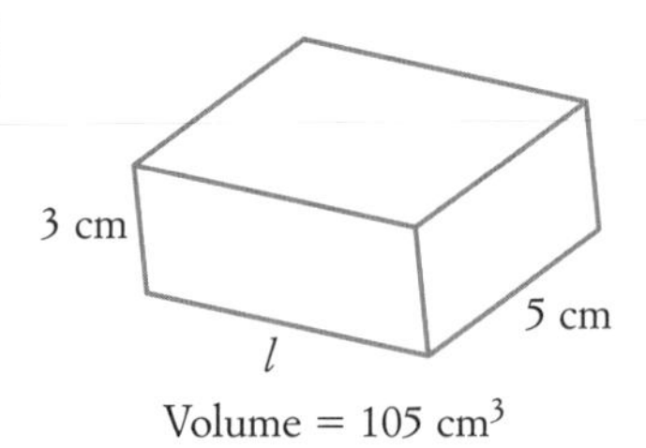

Volume = 105 cm^3

l = ____________

2

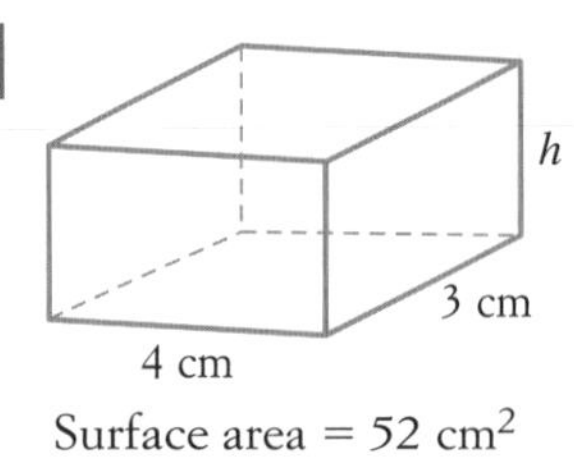

Surface area = 52 cm^2

h = ____________

3

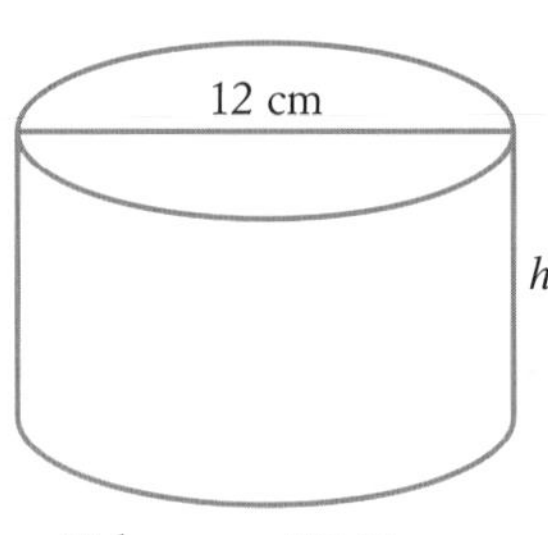

Volume = 452.39 cm

h = ____________

4

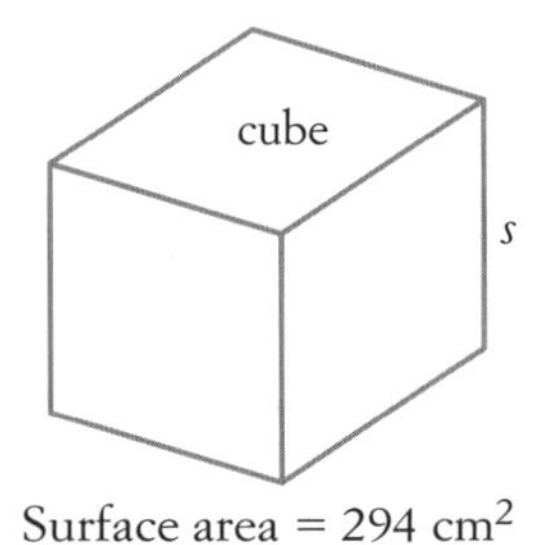

Surface area = 294 cm^2

s = ____________

5

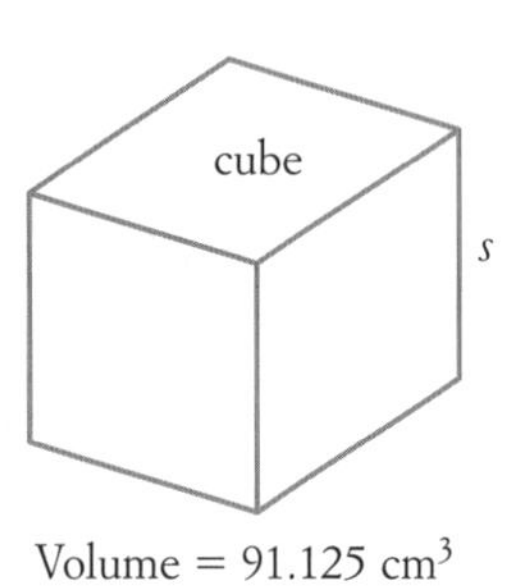

Volume = 91.125 cm^3

s = ____________

6

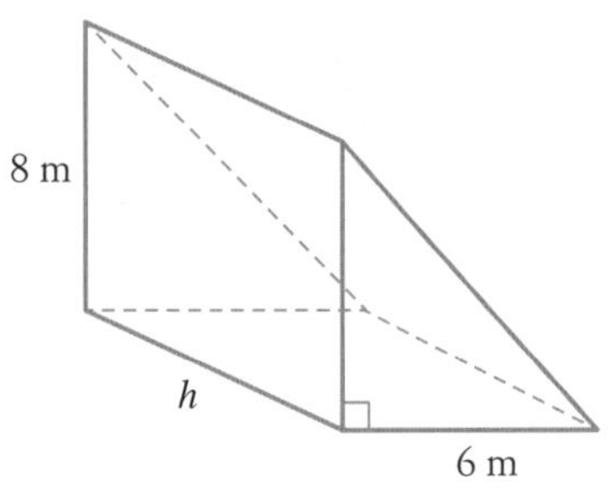

Volume = 288 m^3

h = ____________

7

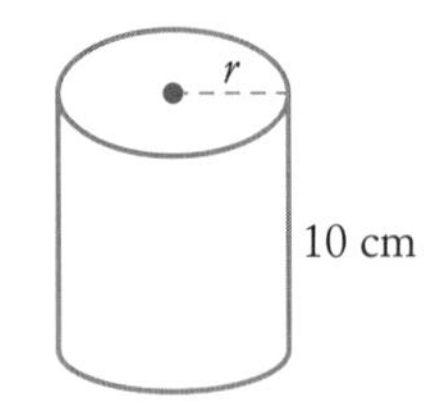

Volume = 502.65 cm^3

r = ____________

8

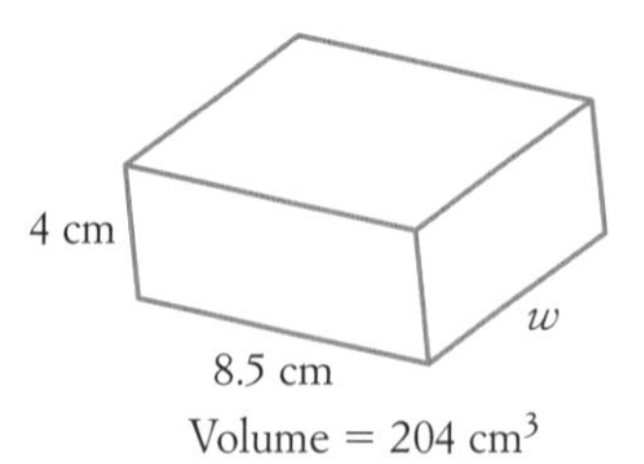

Volume = 204 cm^3

w = ____________

9

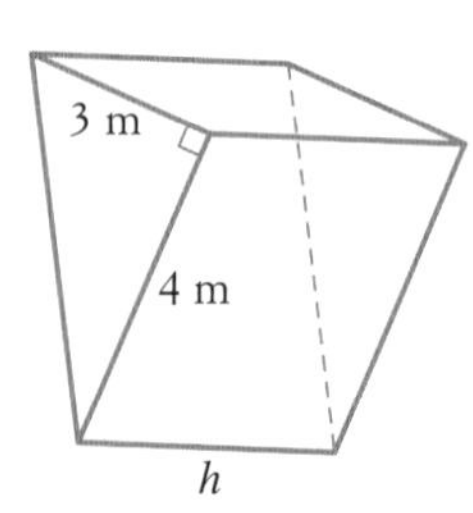

Surface area = 48 m^2

h = ____________

C S F

9780170454551

10

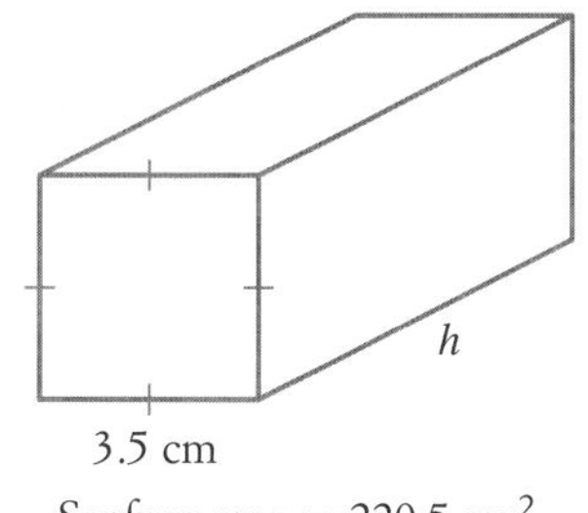

Surface area = 220.5 cm^2

$h =$ ____________

11

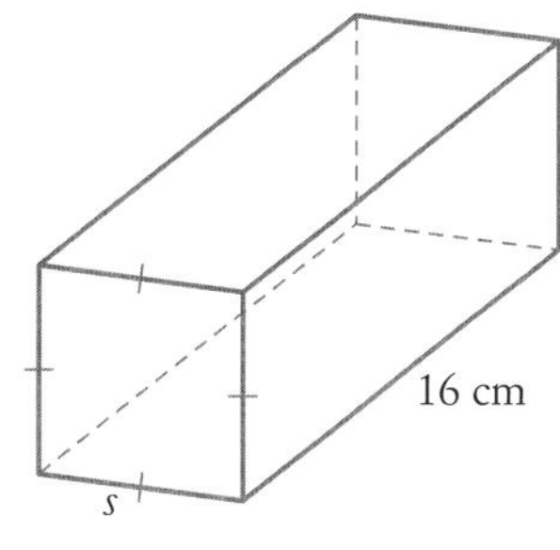

Volume = 231.04 cm^3

$s =$ ____________

12

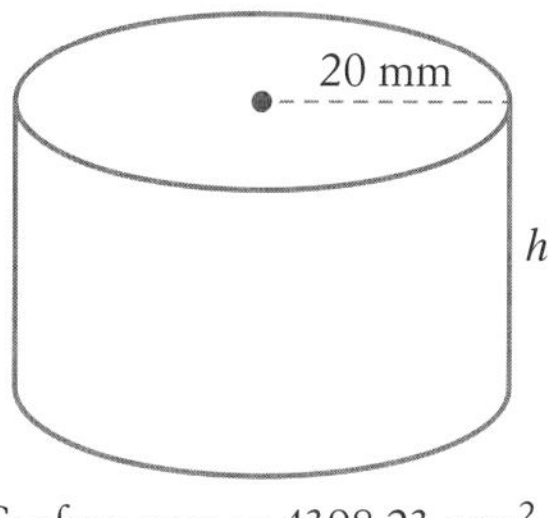

Surface area = 4398.23 mm^2

$h =$ ____________

13

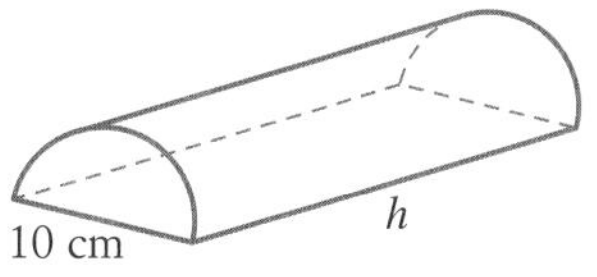

Surface area = 721.24 cm^2

$h =$ ____________

14

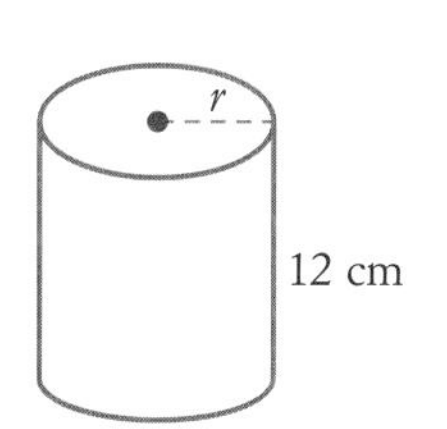

Surface area = 534.07 cm^2

$r =$ ____________

15

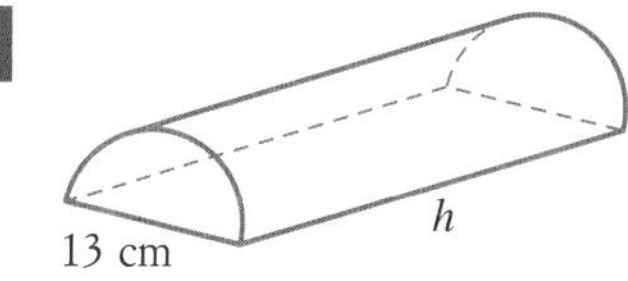

Volume = 1327.32 cm^3

$h =$ ____________

3 SURFACE AREA AND VOLUME CROSSWORD

PS PUZZLE SHEET

ANNULUS	AREA	BASE	CAPACITY
CIRCLE	CIRCUMFERENCE	CLOSED	CONE
CROSS SECTION	CYLINDER	DIAMETER	EXTERNAL
FORMULA	HEIGHT	KILOLITRE	LENGTH
LITRE	NET	OPEN	PI
PRISM	PYRAMID	PYTHAGORAS	QUADRANT
RADIUS	RECTANGULAR	SECTOR	SEMICIRCLE
SOLID	SPHERE	SQUARE	SURFACE
TRIANGULAR	WIDTH		

3 SURFACE AREA

Name:

Due date:

Parent's signature:

Part A	/ 8 marks
Part B	/ 8 marks
Part C	/ 8 marks
Part D	/ 8 marks
Total	/ 32 marks

HEY, I'M MITCH. THIS ASSIGNMENT COVERS MENTAL MATHS, AREA AND SURFACE AREA.

PART A: MENTAL MATHS

Calculators not allowed

1 Solve $\frac{x}{7} = \frac{x-2}{5}$.

2 At a New Year's sale, bags are discounted by 25%. What is the discount price of a bag marked at $150?

3 For this triangle, write as a fraction:

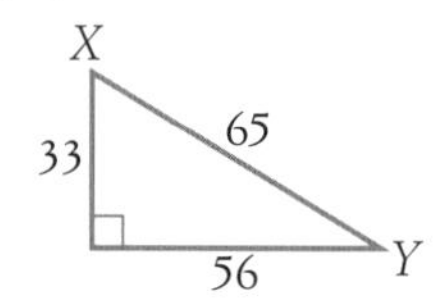

a $\sin X$

b $\cos Y$

c $\tan Y$

4 These dot plots show the ages of students in 2 Taekwondo classes.

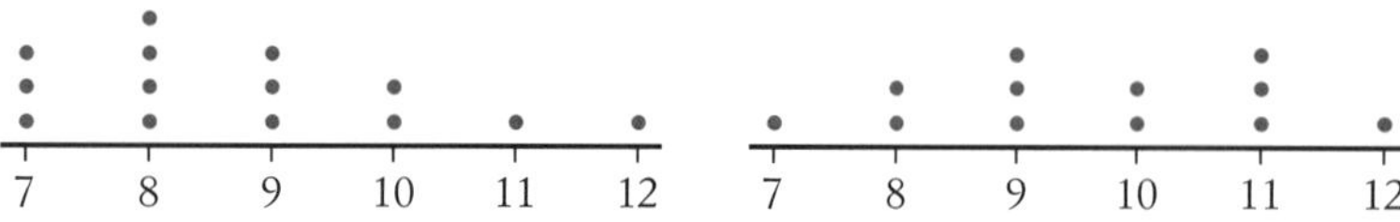

a Which class had more consistent ages?

b Which class had more older students?

5 The scale of a plan is 1 : 2000. What distance in centimetres on the plan would represent a real distance of 50 m?

PART B: REVIEW

1 Find the area of each shape (correct to 2 decimal places for **c** and **e**).

a

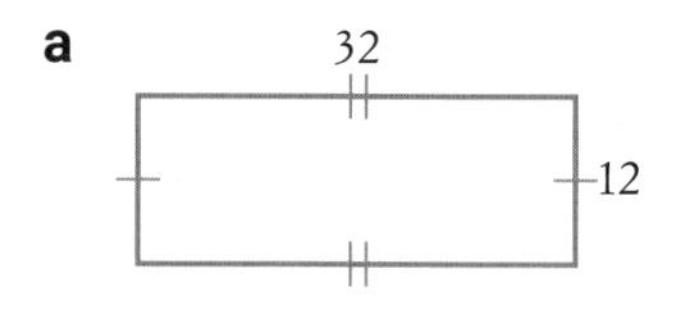

b

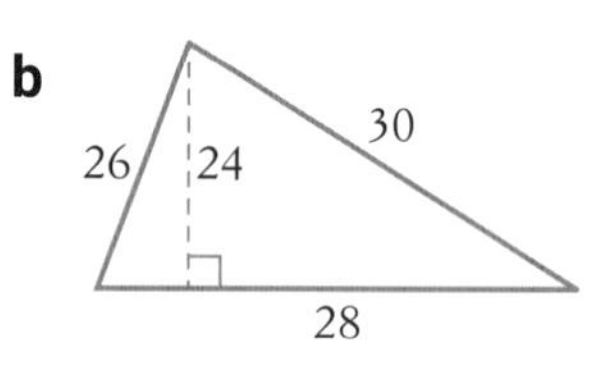

9780170454551

c

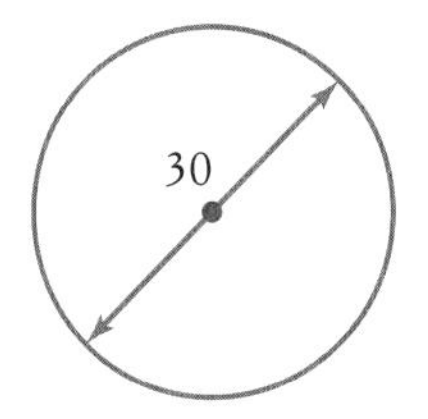

d

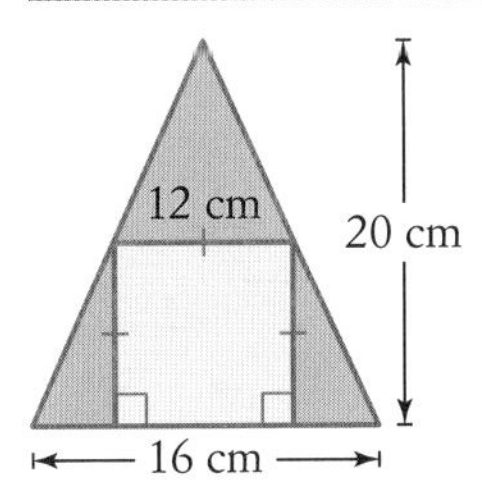

e (2 marks)

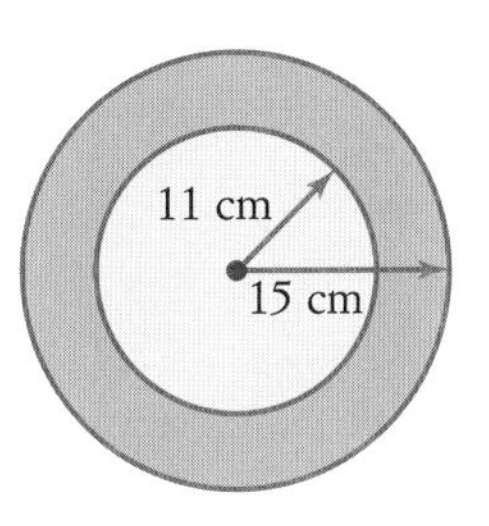

2 Find the value of each variable.

a

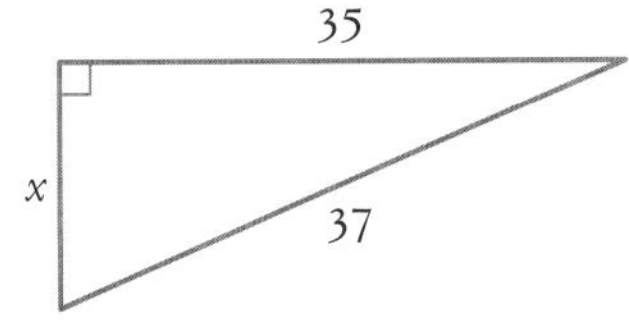

b

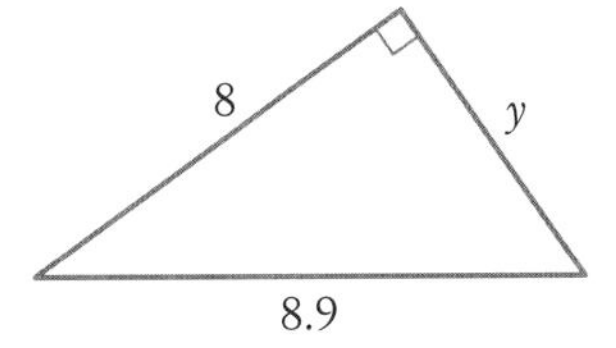

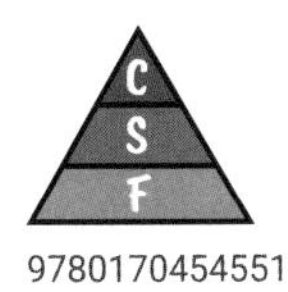

PART B: PRACTICE

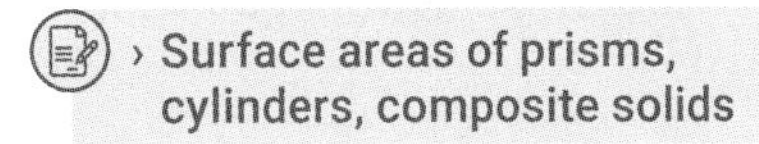

1 (4 marks) Calculate the surface area of each solid (correct to 3 significant figures for **b**). All measurements are in centimetres.

a

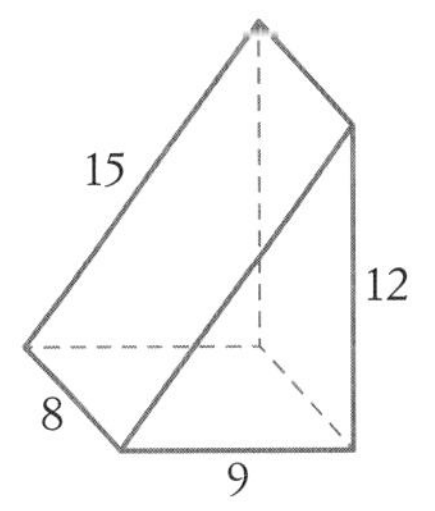

b

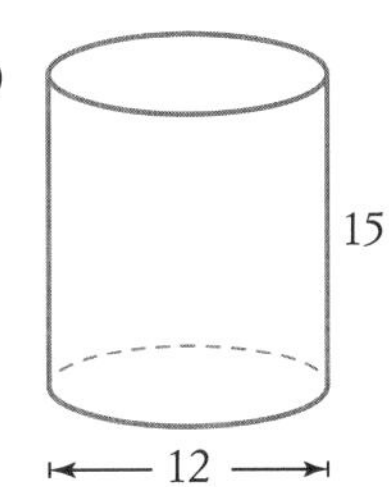

HW HOMEWORK

2 (4 marks) The depth of water in this swimming pool ranges from 1 m to 3 m. Calculate, correct to 2 decimal places:

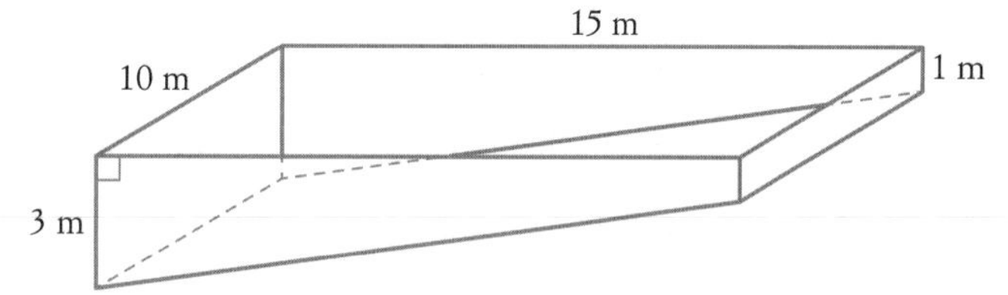

a the area of its slanted floor

b its total surface area

PART D: NUMERACY AND LITERACY

1 What is the **surface area** of a solid?

2 (2 marks) Complete: A prism has the same ______ along its length, and each one is a ______

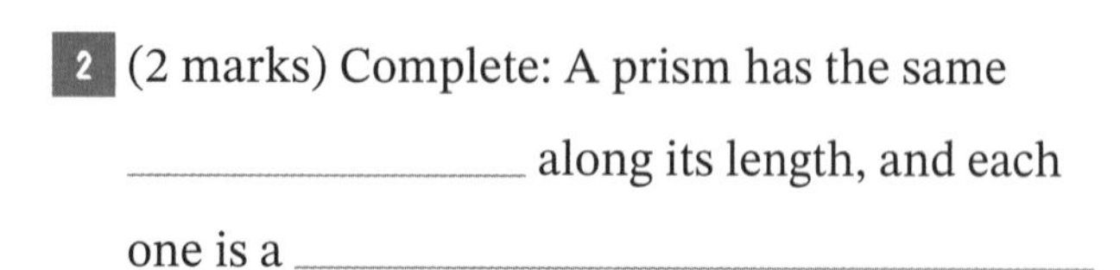

3 The surface area of a cylinder has the formula:

$SA = 2\pi r^2 +$ ______

4 (4 marks) Calculate the surface area of each solid, correct to one decimal place. All measurements are in metres.

a

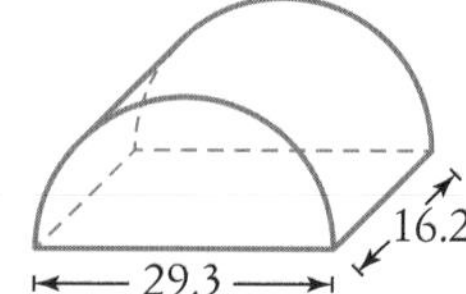

b half-cylinder with open top, one end open

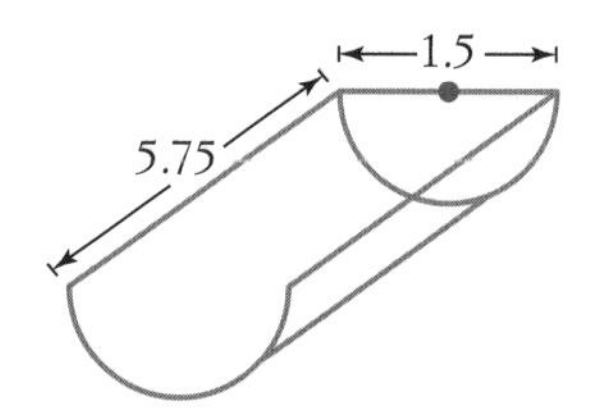

C S F

9780170454551

SURFACE AREA IS THE TOTAL AREA OF ALL FACES AND CURVED SURFACES OF A SOLID.

WS WORKSHEET

Find the surface area of each prism. (Answer questions **5**, **7** and **11** correct to one decimal place.)

1

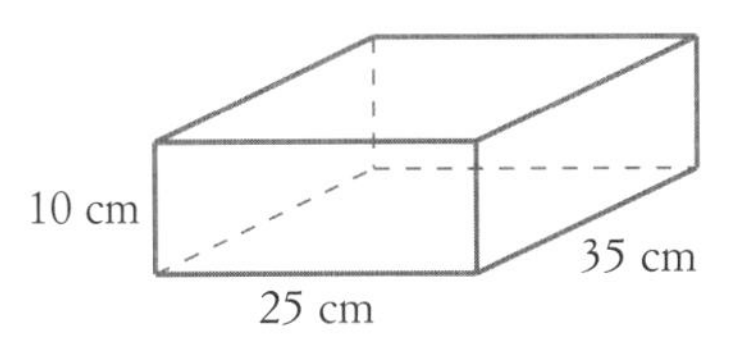

2

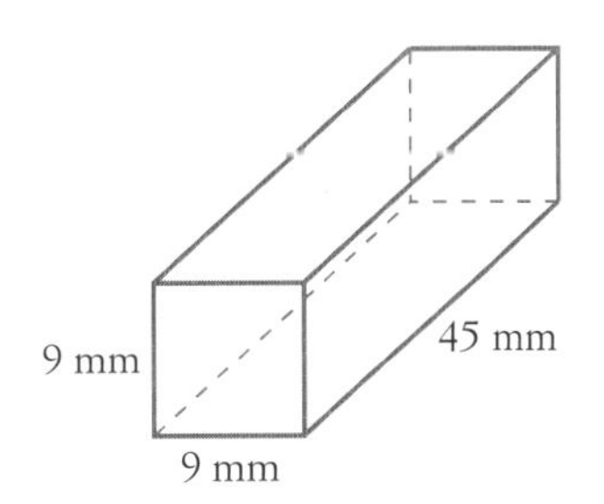

3

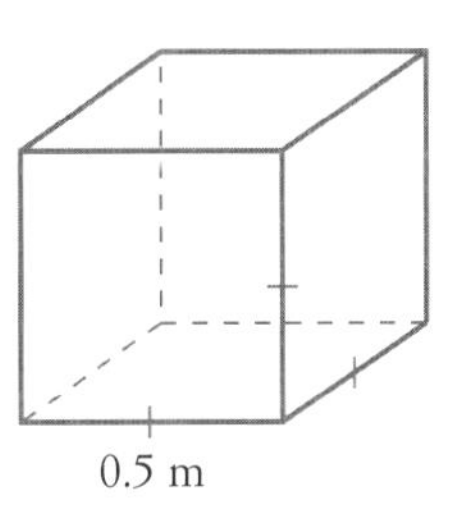

4

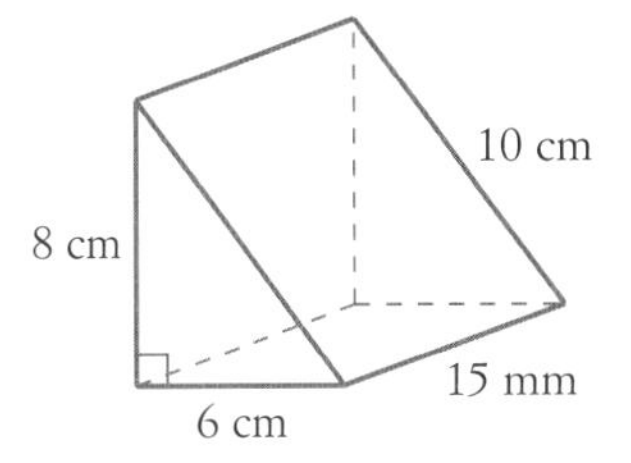

5

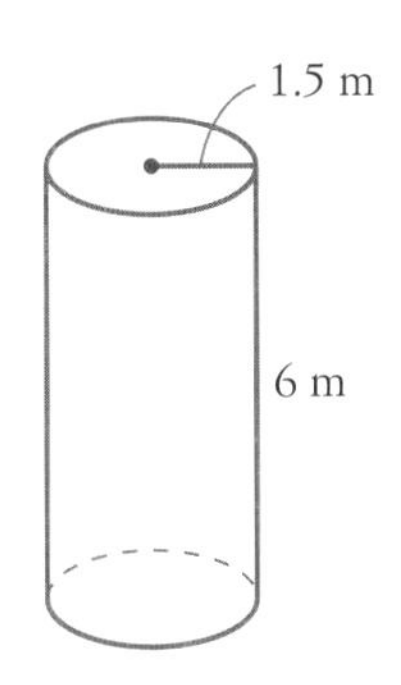

6

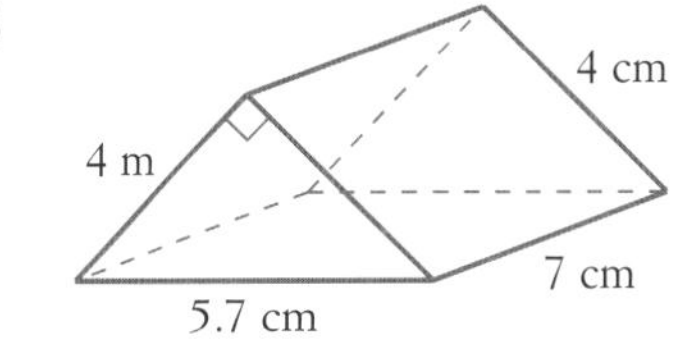

7

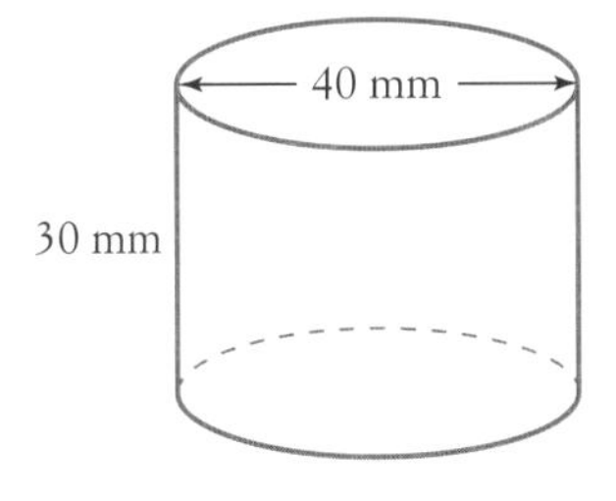

8

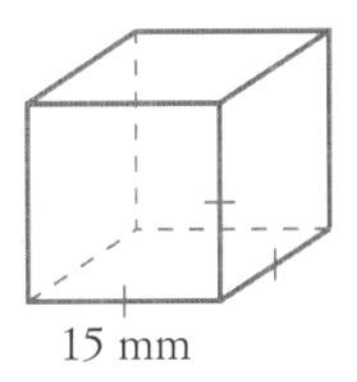

9

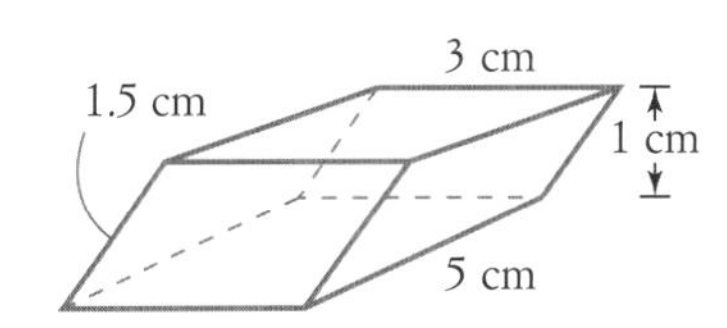

10

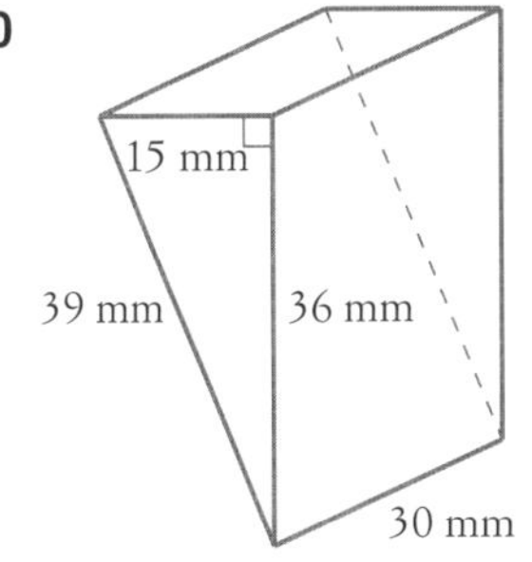

11

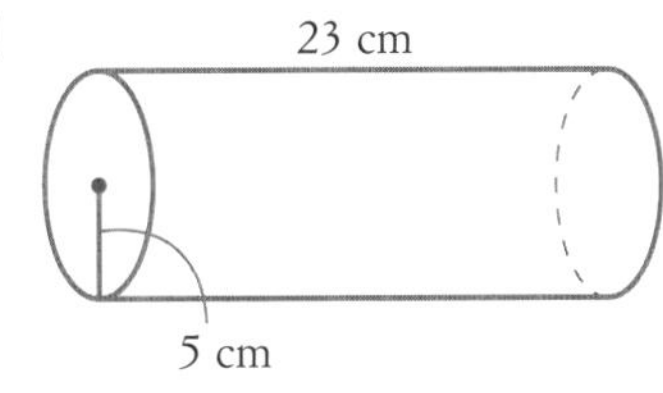

12

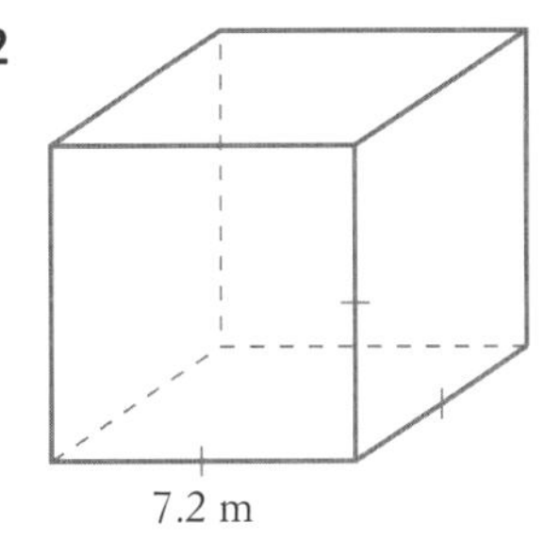

3 VOLUME

THE CORRECT WAY TO SAY M^3 IS 'CUBIC METRES', NOT 'METRES CUBED'

Name:

Due date:

Parent's signature:

Part A	/ 8 marks
Part B	/ 8 marks
Part C	/ 8 marks
Part D	/ 8 marks
Total	/ 32 marks

HW HOMEWORK

PART A: MENTAL MATHS

Calculators not allowed

1 Expand $(x + 3)(x + 12)$.

2 Find 5% of \$3720.

3 Convert 8 m/s to km/h.

4 The rectangle on the left was enlarged to make the rectangle on the right.

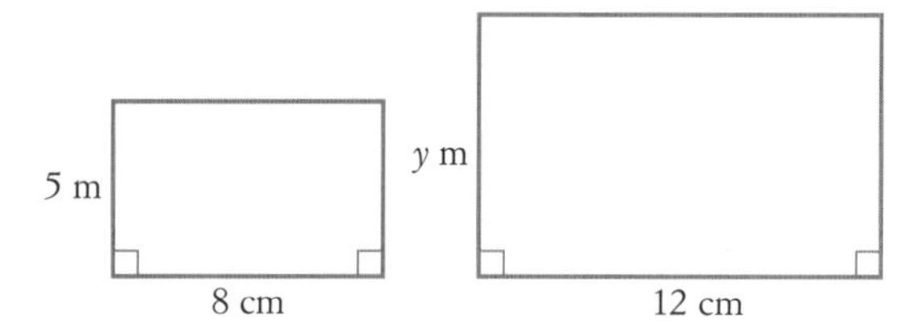

a What is the scale factor?

b Find the value of y.

5 Write the equation with gradient 2 and y-intercept -5.

6 (2 marks) 30 students were surveyed on whether they liked to listen to music or watch TV. 21 students liked music and 18 liked TV, while 5 students did not like either. How many students liked both music and TV?

PART B: REVIEW

Find the surface area of each solid (correct to 2 decimal places for question **3**).

1 (3 marks)

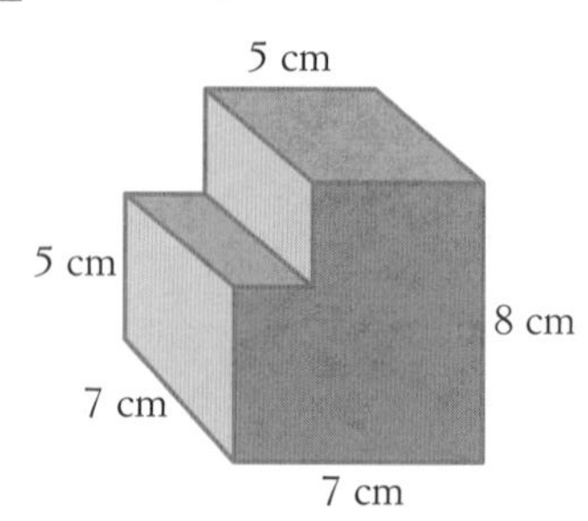

C
S
F

9780170454551

2 (2 marks)

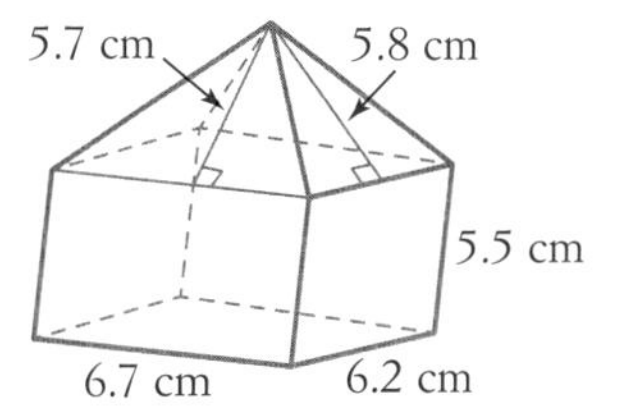

3 (3 marks)

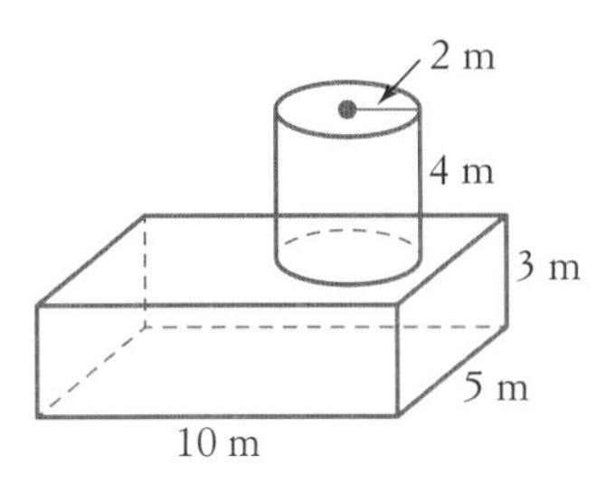

PART C: PRACTICE

› Volumes of prisms, cylinders, composite solids

1 (4 marks) Calculate (correct to the nearest cm^3 for **b**) the volume of each solid.

a

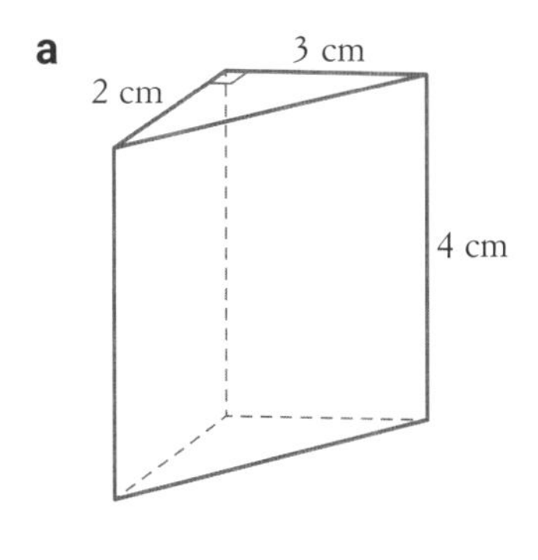

b

20 cm

20 cm

c

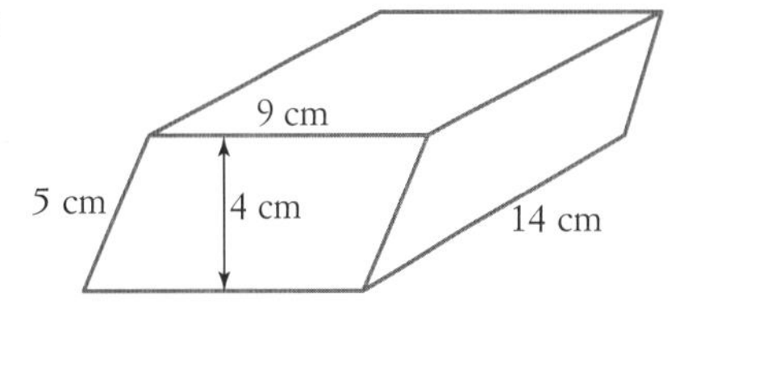

d

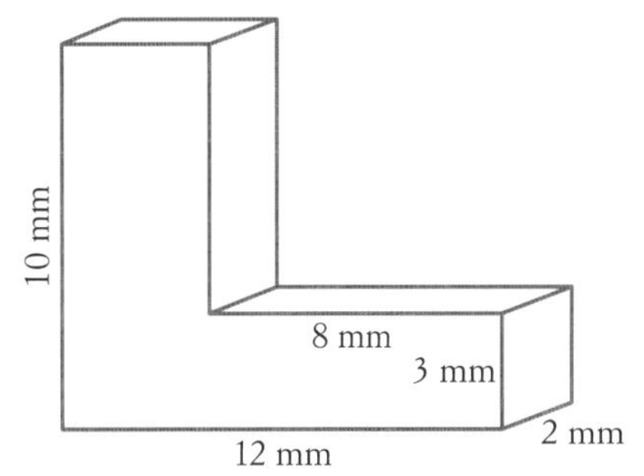

HW HOMEWORK

C S F

2 (4 marks) Calculate (correct to 2 decimal places for **b**) the volume of each solid.

a

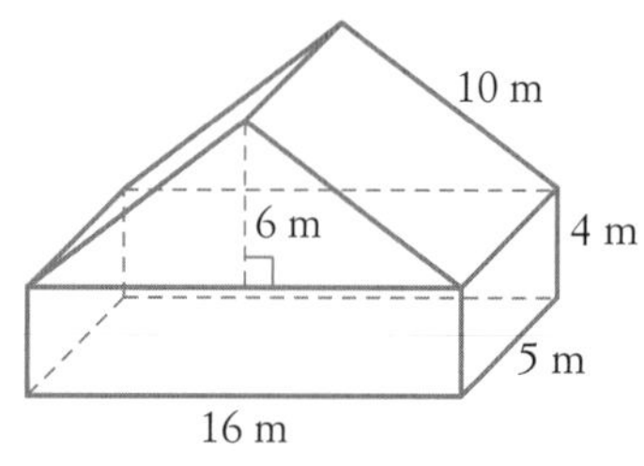

b

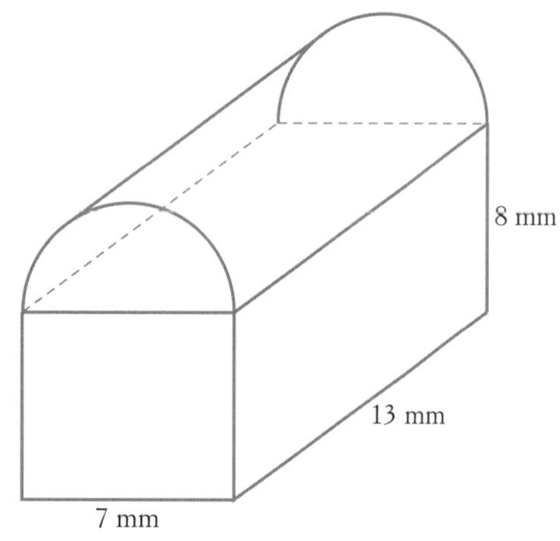

2 (4 marks) Calculate (correct to one decimal place for **a**) the volume of each solid.

a

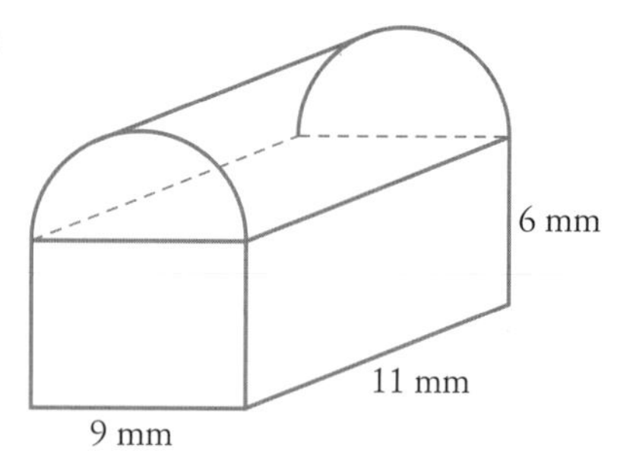

b

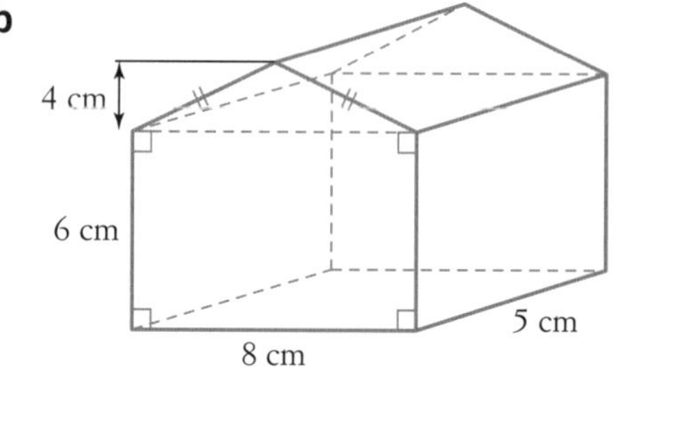

PART D: NUMERACY AND LITERACY

1 Complete:

a The ______________ of a solid is the amount of space it takes up.

b Volume is measured in ______________ units.

c The ______________ of a container is the amount of fluid it holds.

d ‘cm^3’ is the abbreviation for ______________ ______________.

HW HOMEWORK

C S F

9780170454551

STARTUP ASSIGNMENT 4 (4)

PART A: BASIC SKILLS

/ 15 marks

1 Evaluate $2x^2 - 6$ if $x = -1$. ______

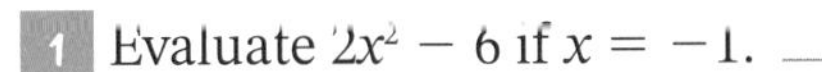

2 Divide \$5100 between Kath and Kim in the ratio 11 : 6.

______

______

______

3 Expand and simplify:

$m(3m - 4) + 2m(5 - m)$ ______

______

______

4 Find the size of one angle in a regular octagon.

______

5 Write 0.000 007 46 in scientific notation.

______

6 Find the median of these numbers:

3, 8, 5, 12, 7, 2

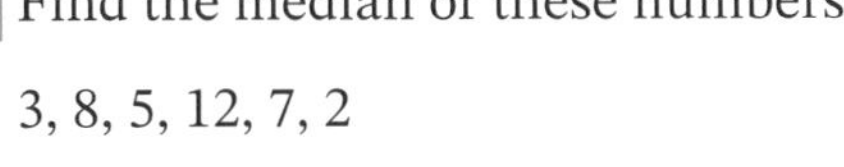

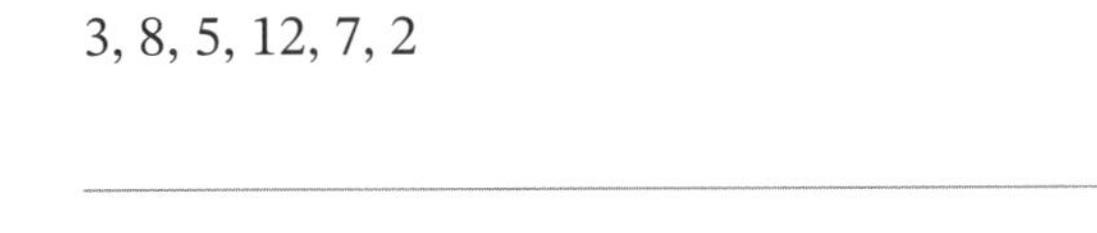

______

7 A rhombus is a square. True or false?

______

8 What is the probability that the next 2 children born are both girls?

______

9 Calculate, correct to 2 decimal places, the surface area of this cylinder.

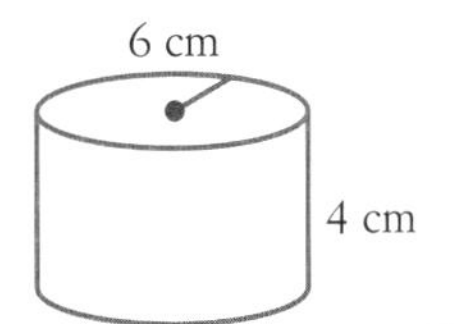

______

10 Solve $\frac{2x - 5}{4} = 7$. ______

______

11 Find the value of k in the diagram below.

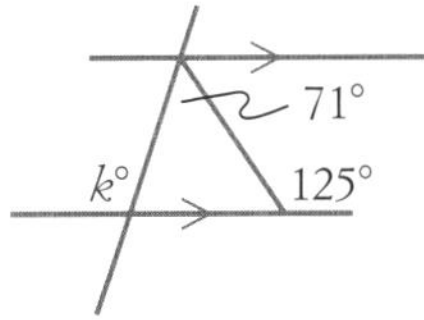

______

12 Graph the line $x = 4$ on a number plane.

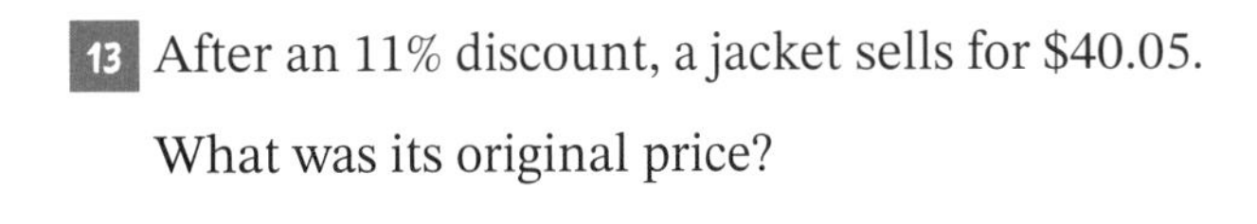

13 After an 11% discount, a jacket sells for \$40.05. What was its original price?

______

______

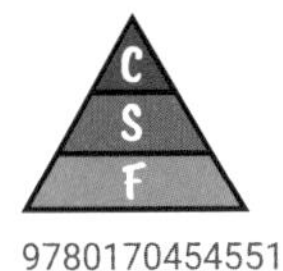

WS WORKSHEET

14 **a** Find the size of $\angle C$ in the diagram below, correct to the nearest degree.

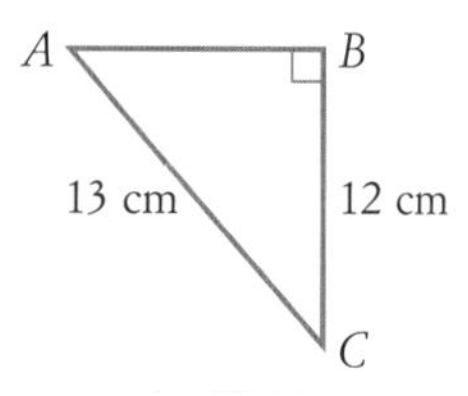

b Calculate the area of the triangle.

PART B: ALGEBRA

/ 25 marks

15 List the factors of:

a 9

b 51

16 Simplify:

a $3m \times 7m^2$

b $\frac{1}{2}x \times 10$

c $4k \times 3k^3$

d $4g \times (-2g)$

e $8p \times 2q$

f $(3u^4)^2$

17 Find the highest common factor of:

a 12 and 20

b $40xy$ and $24y$

18 List the square numbers from 1 to 100.

19 Expand:

a $3(x + 7)$

b $2(5m - n)$

c $4(1 - 3g)$

d $-2(6p + 5)$

e $m(m + 7)$

f $p(3p - 4)$

g $3k(2k - 4)$

h $-5y(3 - y)$

20 Factorise:

a $3x + 12$

b $10m - 20$

c $3p^2 - 5p$

d $-8m + 18$

e $-14y^2 - 8y$

PART C: CHALLENGE

Bonus / 3 marks

You are given 12 coins that look identical, but one is a counterfeit (fake) and weighs *less* than the others.

You are given a balance scale. Can you find the counterfeit coin after only 3 weighings?

C S F

9780170454551

ALGEBRAIC FRACTIONS 4

IF YOU GET STUCK ON THESE, ASK A TEACHER OR FRIEND FOR HELP.

Simplify each expression.

1 $\frac{4x}{10} + \frac{2x}{10}$ ______

2 $\frac{3y}{5} + \frac{2y}{5}$ ______

3 $\frac{3d}{2} + \frac{d}{4}$ ______

4 $\frac{5h}{6} + \frac{h}{2}$ ______

5 $\frac{h}{5} + \frac{4h}{3}$ ______

6 $\frac{9k}{7} + \frac{5k}{2}$ ______

7 $\frac{4x}{3} + \frac{5x}{8}$ ______

8 $\frac{4}{2x} + \frac{6}{2x}$ ______

9 $\frac{11}{3d} + \frac{7}{d}$ ______

10 $\frac{7y}{2} - \frac{4y}{2}$ ______

11 $\frac{10p}{3} - \frac{4p}{3}$ ______

12 $\frac{10r}{4} - \frac{3r}{2}$ ______

13 $\frac{6e}{10} - \frac{e}{5}$ ______

14 $\frac{13d}{6} - \frac{3d}{4}$ ______

15 $\frac{9p}{7} - \frac{p}{3}$ ______

16 $\frac{14}{5d} - \frac{10}{d}$ ______

17 $\frac{4r}{3} + \frac{5r}{3} - \frac{6r}{3}$ ______

18 $\frac{g}{2} + \frac{3g}{4} - \frac{6g}{4}$ ______

19 $\frac{2e}{5} \times \frac{5y}{6}$ ______

20 $\frac{4r}{3} \times \frac{6t}{3}$ ______

21 $\frac{9y}{2} \times 10$ ______

22 $8 \times \frac{5g}{2}$ ______

23 $\frac{6f}{4} \times \frac{20}{3}$ ______

24 $\frac{25h}{12} \times \frac{16g}{10}$ ______

25 $\frac{22d}{10e} \times \frac{15e}{18}$ ______

26 $\frac{14r}{20} \times \frac{6r}{7}$ ______

27 $\frac{10e}{14d} \times \frac{10e}{14d}$ ______

28 $\frac{5d}{2} \div \frac{25}{3}$ ______

29 $\frac{20a}{6} \div \frac{4a}{3}$ ______

30 $\frac{30f}{4} \div \frac{10}{7e}$ ______

31 $\frac{12ad}{9c} \div \frac{3d}{4}$ ______

32 $\frac{16xy}{10} \div \frac{6x}{15y}$ ______

33 $\frac{24}{6g} \div \frac{8g}{h}$ ______

34 $\frac{9cd}{4e} \div \frac{15c}{14e}$ ______

35 $\frac{36mn}{5} \div 4$ ______

36 $\frac{28h}{16f} \div 6$ ______

WS WORKSHEET

9780170454551

4 ALGEBRA CROSSWORD

PS PUZZLE SHEET

Write the answers to the clues and complete this puzzle.

9780170454551

Clues across

3 The answer to a multiplication

8 An algebraic expression with 2 terms, for example $x + 9$, $2y - 12$.

10 The number at the top of a fraction

13 ( )

14 The plural of index

16 $\frac{a^m}{a^n} = a^{m-n}$ is an __________ law

18 In 5^3, 5 is called the __________

19 The __________ of 12 are 1, 2, 3, 4, 6 and 12

Clues down

1 In $x^2 - 2x + 9$, 9 is called the __________ term

2 $x^2 - 2x + 9$ is an example of a __________ expression

3 In 5^3, 3 is called the __________

4 When expanding an algebraic expression, multiply each __________ inside the brackets by the __________ outside the brackets

5 To take out the HCF of an expression and write it with brackets

6 The H in HCF

7 The number in front of a variable, such as -4 in $-4x$

9 $\frac{x}{2}$ is an example of an __________ fraction

11 Raising a number to a power of -1 gives its __________

12 Abbreviation for 'highest common factor'

15 To remove the brackets of an algebraic expression

17 The number at the bottom of a fraction

PS PUZZLE SHEET

4 INDEX LAWS

Name:

Due date:

Parent's signature:

Part A	/ 8 marks
Part B	/ 8 marks
Part C	/ 8 marks
Part D	/ 8 marks
Total	/ 32 marks

HW HOMEWORK

PART A: MENTAL MATHS

Calculators not allowed

1 (2 marks) Write 2^4 in expanded form, then evaluate it.

2 Round 0.084 55 to 2 significant figures.

3 Express 0.0008 in scientific notation

4 Find, as a surd, the distance between the points $A(6, 2)$ and $B(-2, 5)$ on the number plane.

5 A card is drawn randomly from a normal deck of cards. Find the probability that it is a King or a Queen card.

6 For these numbers, find:

8 6 12 4 10

a the mean

b the range

PART B: REVIEW

1 Simplify each expression.

a $x \times y \times x \times y \times x$

b $5 \times p \times t \times p \times 5 \times p \times t$

2 Evaluate each expression.

a $3^2 \times 3^4$

b $10^6 \div 10$

c $(2^3)^2$

d 18^0

3 Find the missing power.

a $8 = 2^{\square}$

b $81 = 3^{\square}$

C S F

9780170454551

PART C: PRACTICE

› Index laws
› Zeros and negative indices

Simplify each expression.

1 $2x^2y^2 \times 7x^3y^8$ ______

2 $\left(\frac{p^5}{9y}\right)^2$ ______

3 $-2n^0$ ______

4 $(-2n)^6$ ______

5 x^{-5} ______

6 $(11p)^{-3}$ ______

7 $\left(\frac{1}{10p}\right)^{-1}$ ______

8 $\left(\frac{2}{5a}\right)^{-2}$ ______

PART D: NUMERACY AND LITERACY

1 **a** What is another name for index?

______

b What does an index show?

______

2 Complete each index law.

a $\frac{a^m}{a^n}$ ______

b $(ab)^n =$ ______

c $a^{-1} =$ ______

3 Simplify each expression.

a $\frac{2p^0}{3} =$ ______

b $\left(\frac{2p}{3}\right)^0 =$ ______

4 Write $\frac{8}{p^3}$ using a negative index.

______

HW HOMEWORK

C
S
F

4 ALGEBRAIC FRACTIONS

Name:

Due date:

Parent's signature:

Part	Marks
Part A	/ 8 marks
Part B	/ 8 marks
Part C	/ 8 marks
Part D	/ 8 marks
Total	/ 32 marks

PART A: MENTAL MATHS

Calculators not allowed

1 Find a simplified algebraic expression for the volume of this rectangular prism.

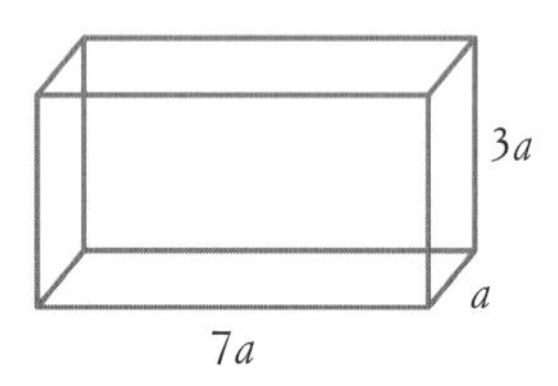

2 (2 marks) Expand $(10x - 2y)(10x + 2y)$.

3 Write the equation of the line that has a gradient of -2 and a y-intercept of 3.

4 A card is chosen at random from a set of cards numbered 0 to 9. Find the probability of selecting:

a 7

b a number less than 5

c a number 6 or more

5 Write $\tan\theta$ for this triangle.

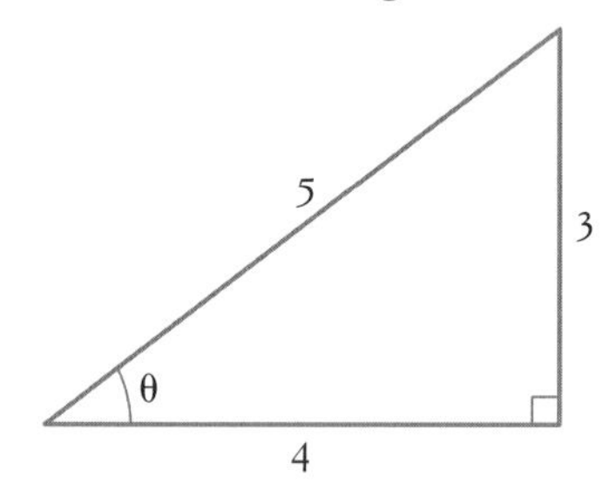

PART B: REVIEW

1 Evaluate without a calculator:

a $\frac{2}{3} - \frac{3}{7}$

b $\frac{4}{9} + \frac{5}{6}$

c $\frac{6}{7} \times \frac{5}{8}$

d $\frac{5}{9} \div \frac{5}{6}$

2 Simplify each expression.

a $\frac{3m^2n^5}{18mn^2}$

b $2t^{-1}$

c $(5x^3y^2)^2$

d $(-3ab^2)^4$

9780170454551

PART C: PRACTICE

› Algebraic fractions

Simplify each expression.

1 $\frac{4x}{15}+\frac{3x}{10}$ ____________________

2 $\frac{y}{16}-\frac{y}{24}$ ____________________

3 $\frac{m}{5}-\frac{m}{7}$ ____________________

4 $\frac{y}{4}+\frac{2y}{5}$ ____________________

5 $\frac{20y}{3x}\times\frac{9y}{5x}$ ____________________

6 $\frac{4mn}{9}\times\frac{3m}{16n}$ ____________________

7 $\frac{8a}{7}\div\frac{40a^2}{3p}$ ____________________

____

8 $\frac{6mn}{5x}\times\frac{4x}{n}\div18mn$ ____________________

____

PART D: NUMERACY AND LITERACY

1 (5 marks) Complete:

a To add or subtract fractions, convert them (if needed) so that they will have the same ____________________, then add or subtract the ____________________.

b To multiply fractions, cancel any common factors, then multiply the ____________________ and ____________________ separately.

c To divide by a fraction $\frac{a}{b}$, multiply by its ____________________ $\frac{b}{a}$.

2 Simplify $\frac{3x-15}{7}\times\frac{14x}{x^2-5x}$

____

____

____

3 Explain in words how to simplify:

a 3^{-2} ____________________

____

____

b 3^0 ____________________

____

____

HW HOMEWORK

C S F

4 EXPANDING AND FACTORISING

Name:

Due date:

Parent's signature:

Part A	/ 8 marks
Part B	/ 8 marks
Part C	/ 8 marks
Part D	/ 8 marks
Total	/ 32 marks

PART A: MENTAL MATHS

Calculators not allowed

1 Expand and simplify:

a $5y(y - 2) - (6 - y)$

b $-3x^2(9 + x) + 10x^3$

2 Write the formula for the area of a circle with radius r.

3 Factorise $10y - 4y^2$.

4 Round 0.8109 correct to 3 significant figures.

5 The ages of the members of the Wang family are:

13 2 24 59 21 48 30 31

Find the median.

6 Raina has a probability of 78% of winning a game. What is the probability that she will not win a game?

7 At the concert, the ratio of boys to girls is 2 : 5. If there are 10 boys, how many girls are there?

PART B: REVIEW

1 Simplify each expression.

a $21x^6 \div 7x^2$

b $5a^4 \times 6a^5$

c $\dfrac{m}{36} \div \dfrac{n}{4}$

d $\dfrac{9}{p} \times \dfrac{7p}{54}$

2 Expand $-8y(4y + 13)$.

9780170454551

3 Find 2 numbers whose:

a product is 36 and sum is 13

b product is -72 and sum is 1

4 Factorise $-24b + 16$

PART C: PRACTICE

1 Expand and simplify each expression.

a $4xy(5y - 8x)$

b $-3a(6 + 2a)$

c $2p(p + 4p^2) - p^2(1 - 2p)$

d $5a(b + 6) - 6b(9 - 3a)$

2 Factorise each expression fully.

a $36x + 24x^2$

b $-8xy + 2y^2$

c $3(p - q) + p(p - q)$

d $16a^3b + 8a^2b$

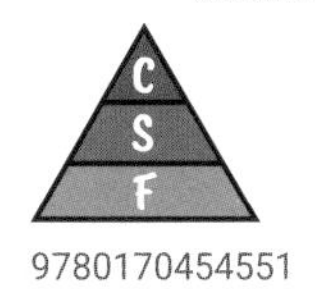

PART D: NUMERACY AND LITERACY

1 Complete each expansion.

a $x(y + z) =$ ____

b $k(r - s) =$ ____

2 a What does HCF stand for?

b What is the HCF of $8mn - 2n^2$?

c Factorise $8mn - 2n^2$.

3 Complete: When factorising an expression, we must ____ each term by the HCF and write the answers inside the brackets.

4 (2 marks) Expand $(a - b)(a + b) - (a + b)^2$.

HW HOMEWORK

4 BINOMIAL PRODUCTS

Name:

Due date:

Parent's signature:

Part A	/ 8 marks
Part B	/ 8 marks
Part C	/ 8 marks
Part D	/ 8 marks
Total	/ 32 marks

PART A: MENTAL MATHS

Calculators not allowed

1 Solve $x^2 = 16$.

2 Find the value of S in the formula $M = \frac{kS}{5}$ if $M = 40$ and $k = 20$.

3 Find x.

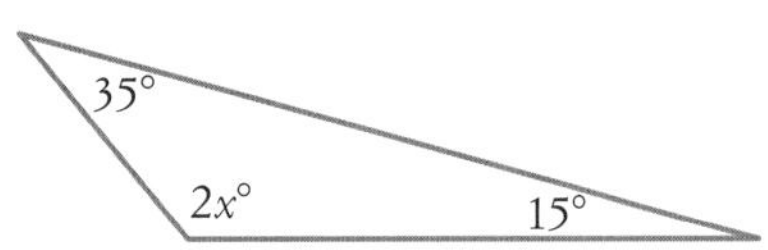

4 For this shape, find:

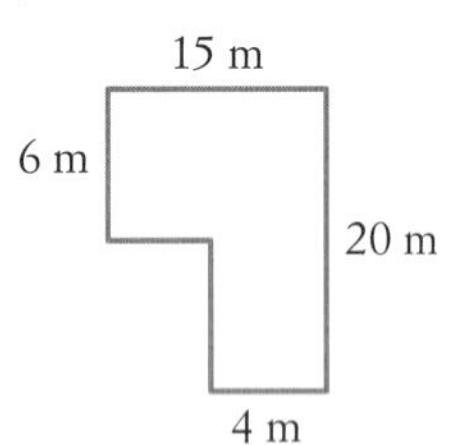

a its perimeter

b its area

5 (3 marks) 2 dice are rolled and their sum calculated. Find the probability of rolling a sum:

a of at least 11

b between 4 and 8

PART B: REVIEW

1 (6 marks) Expand each expression.

a $3xy(2xy - 2)$

b $-4a(4a^2 + 3b)$

c $4(2 - m) - m(6 + 3m)$

d $4a(4a - 7) - 6(4a - 7)$

C S F

9780170454551

2 Factorise each expression.

a $8xy^2 + 4xy + 44x^2y$

b $-3a^5 - 6a^4 + a$

PART C: PRACTICE

1 (6 marks) Expand each expression.

a $(x + 4)(x - 3)$

b $2y(y + 6) - 10(y + 6)$

c $(2n - 7)(n + 5)$

2 Factorise each expression.

a $x^2 + 2x - 63$

b $y^2 - 4y - 32$

PART D: NUMERACY AND LITERACY

1 What is a binomial expression? Give an example of one.

2 What is the product of 9 and -3?

3 What is a quadratic expression? Give an example of one.

4 Find 2 numbers that have a sum of -5 and a product of -14.

5 Factorise each expression.

a $25x^2 + 5xy$

b $-81p + 9p^2$

c $3ab(b - 9a) - 9a^2b$

d $3x(2x - 5) - 5(2x - 5)$

HW HOMEWORK

C S F

4 MIXED EXPANSIONS

Expand each expression. Mixed answers are given below.

1 $(b+7)(b+5)$	**2** $2b(b-18)$	**3** $(3b+1)^2$
4 $(b+8)(b-8)$	**5** $(b-3)(b+9)$	**6** $(b-1)(b+1)$
7 $(b+3)(b-3)$	**8** $(b-c)(c-2)$	**9** $(b+c)(b+1)$
10 $(b+8)^2$	**11** $(x-4)(x-8)$	**12** $(x-7)^2$
13 $(4x+5)(x-2)$	**14** $(2x+9)(2x-5)$	**15** $(x-5)^2$
16 $-7(2x-3)$	**17** $(6x-2)(6x+2)$	**18** $(2x-6)^2$
19 $(2x+1)(2x-7)$	**20** $(2x-3)(2x+3)$	**21** $-m(m+7)$
22 $(m+7)^2$	**23** $(2m+4)(m-7)$	**24** $(3m-4)(2m+3)$
25 $(4m+1)(4m-1)$	**26** $(m+3)(m+4)$	**27** $(m+5)(m-1)$
28 $5(m+1)-(m+1)$	**29** $(m+5)(m-2)$	**30** $(3m-2)^2$

Mixed answers

$4x^2+8x-45$	$6m^2+m-12$	$4x^2-9$	$b^2+16b+64$	$4x^2-3x-10$
$m^2+3m-10$	$-14x+21$	$9m^2-12m+4$	$b^2+b+bc+c$	$2m^2-10m-28$
b^2-9	$2b^2-36b$	$16m^2-1$	m^2+4m-5	$4m+4$
$m^2+7m+12$	$4x^2-12x-7$	$b^2+6b-27$	$x^2-12x+32$	$-m^2-7m$
$b^2+12b+35$	$36x^2-4$	$4x^2-24x+36$	$bc-2b-c^2+2c$	$9b^2+6b+1$
b^2-64	$m^2+14m+49$	$x^2-10x+25$	b^2-1	$x^2-14x+49$

STARTUP ASSIGNMENT 5

LET'S GET READY FOR THE COMPARING DATA TOPIC BY REVISING OUR STATISTICS SKILLS.

WS WORKSHEET

PART A: BASIC SKILLS

/ 15 marks

1 Simplify $(5a)^3$. ______

2 Which size pack of ice cream below gives better value for money?

A 750 g for $5.20

B 1.25 kg for $8.60

______

3 Find the median of these numbers:

8, 8, 7, 4, 5, 4, 3, 6, 6, 7 ______

4 Simplify $\frac{6x - 4xy}{2x}$. ______

5 Name this solid.

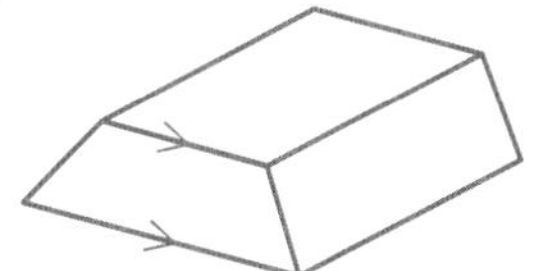

______

6

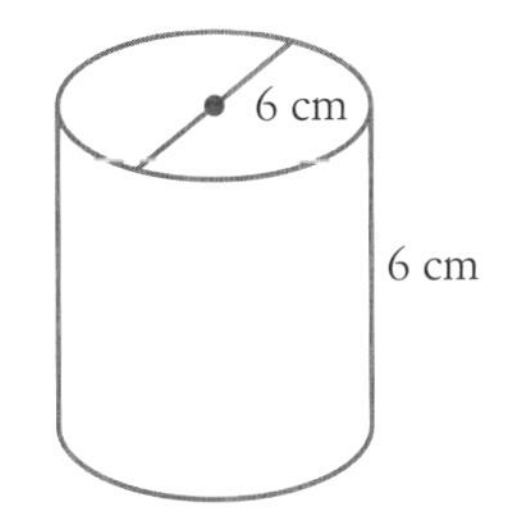

For this cylinder, find to 2 decimal places:

a its volume ______

b its surface area. ______

7 Write 62.5% as a fraction. ______

8 Find k if 8 more than triple k is 2 less than 4 times k. ______

______

______

9 Complete: 3.6 hours = 3 h ____ min

10 Solve $\frac{d}{5} = \frac{16}{3}$. ______

______

11 What is the y-intercept of the graph of $y = -2x$? ______

12 The value of a house increased from \$440 890 to \$491 600. Calculate the percentage increase in value, correct to one decimal place.

______

13 A rectangle has a perimeter of 450 cm. Find its length and width if they are in the ratio 5 : 4.

______

______

14 A car has a fuel consumption of 8.1 L/100 km. How many whole kilometres can it travel on 45.4 L of fuel? ______

______

PART B: STATISTICS

/ 25 marks

15 This dot plot displays the results of a survey about the number of TVs in homes.

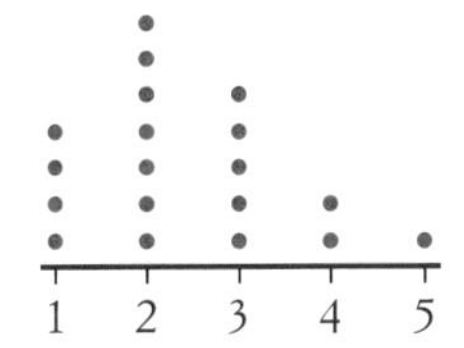

a How many homes were surveyed?

______

b Mode = ______

c Range = ______

9780170454551

d Mean (to 2 decimal places) = ____________

e Median = ____________

f Draw a frequency polygon for the data.

16 What is the difference between a **sample** and a **census**?

____

____

____

____

17 This stem-and-leaf plot shows masses (in kilograms) of a group of students.

Stem	Leaf
5	2 4 5 7 9
6	0 3 3 4 4 4 6 7 8 8
7	0 1 1 2 3 6

Complete the following:

a Mode = ____________

b Range = ____________

c Median = ____________

d How many students weighed less than 65 kg?

____

e Mean = ____________ (to 2 decimal places)

18 This data is from a survey about the number of people living in individual houses.

4 2 1 4 3 5 8 6 3 5

3 1 2 5 4 3 2 7 6 4

5 3 2 4 4 2 4 4 5 3

a Complete the cumulative frequency table.

Score	f	cf
1		
2		
3		
4		
5		
6		
7		
8		

b Mode = ____________

c Median = ____________

d Range = ____________

e How many houses had 3 or fewer people living in them? ____________

PART C: CHALLENGE

Bonus / 3 marks

64 players entered a tennis knockout competition. After each round, only the winners continue playing; the losers are 'knocked out'. How many matches are required to find the winner of the competition? How many rounds are required?

____

C S F

9780170454551

INTERQUARTILE RANGE 5

For each set of data, find the median (Q_2), lower quartile (Q_1), upper quartile (Q_3) and interquartile range ($Q_3 - Q_1$).

		Median	Q_1	Q_3	Interquartile range
1	8, 11, 11, 15, 17, 20, 21				
2	10, 10, 16, 16, 20, 25, 26, 28				
3	3, 4, 5, 8, 11, 11, 14, 14, 14, 19, 20				
4	0, 1, 1, 1, 4, 5, 8, 10, 12				
5	18, 19, 20, 20, 20, 25, 27, 31, 34, 40				
6	7, 15, 19, 25, 29, 31, 40, 40, 50, 55				
7	54, 54, 60, 64, 64, 64, 68, 68, 70				
8	2, 1, 5, 3, 4, 5, 2				
9	1, 84, 86, 88, 90, 83				
10	10, 20, 30, 40, 50				
11	5, 4, 6, 3, 2, 4, 6, 9, 4, 7, 3, 2, 3				
12	1, 0, 2, 3, 0, 1, 4, 2, 3, 0, 1, 1, 5, 4, 3, 2				
13	7, 6, 9, 9, 8, 7, 7, 3				
14	9, 9, 9, 9, 9, 10				
15	8, 10, 12, 7, 8, 10, 9, 8, 10, 8, 11				
16	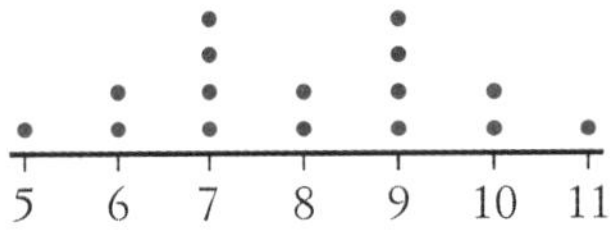 5 6 7 8 9 10 11				
17	Stem | Leaf 2 | 0 2 2 5 7 3 | 1 3 4 4 4 9 4 | 2 6				
18	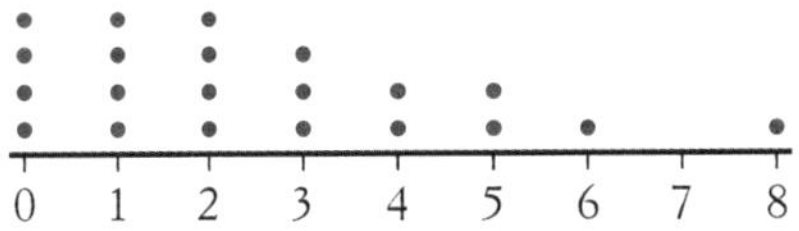 0 1 2 3 4 5 6 7 8				

5 BOX-AND-WHISKER PLOTS

1 The box-and-whisker plots below illustrate the annual road fatalities in Australia over a 10-year period.

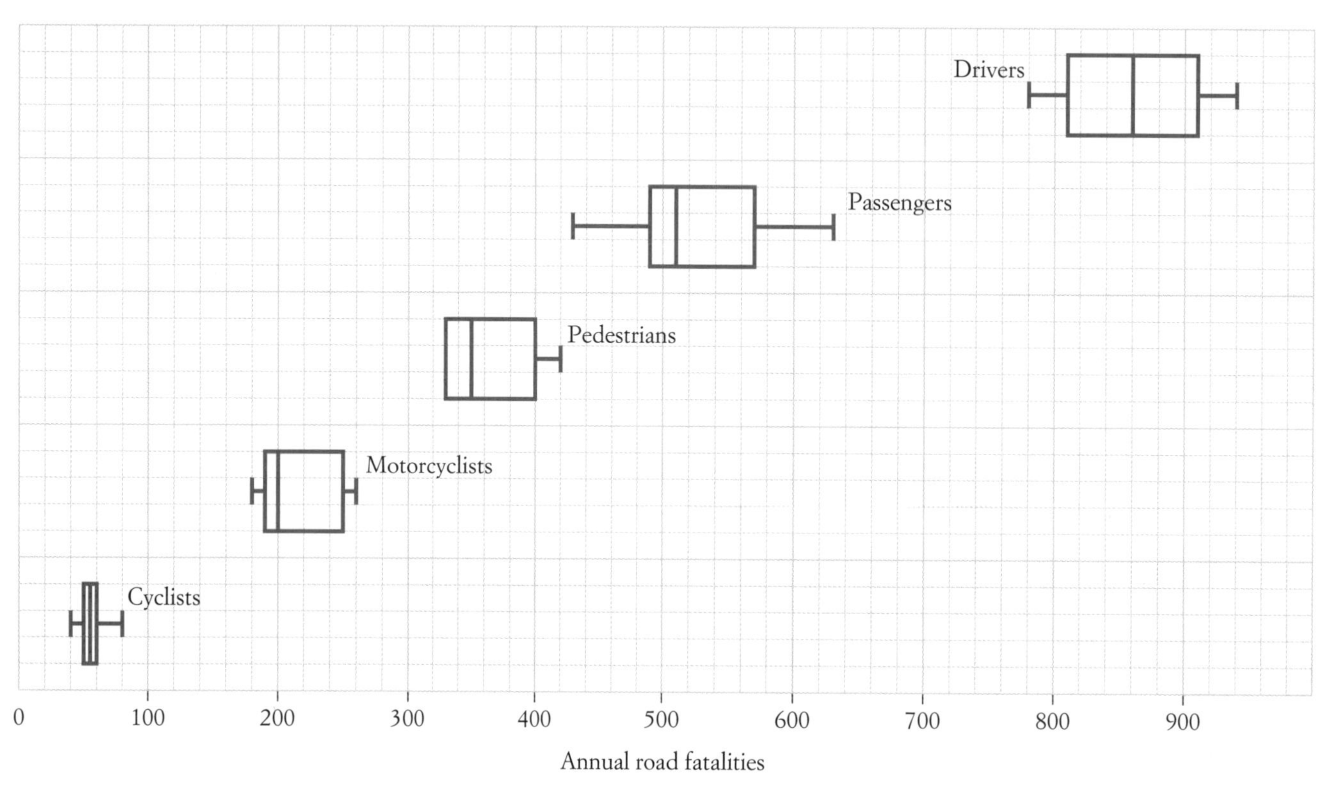

a Which group of road users had:

i the most fatalities? ______

ii the greatest range of data? ______

iii the greatest interquartile range? ______

iv a symmetrical distribution of data? ______

b What was the median number of pedestrian fatalities? ______

c What was the range for the motorcyclist fatalities? ______

d What was the interquartile range for passenger fatalities? ______

e Is the distribution for the pedestrian data skewed positively or negatively? ______

f Which score for the cyclist data was an outlier? ______

g The box-and-whisker-plot for the cyclist data is the shortest. What does this mean about the annual fatality rate for cyclists? ______

2 The half-yearly exam marks for two Year 12 Maths classes are listed below:

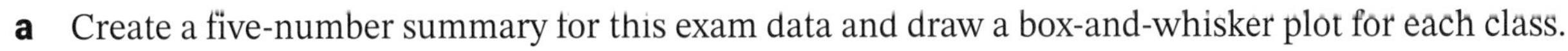

Class 12J: 56 67 70 40 48 84 60 59 58 65 44

48 45 53 28 64 46 77 50 54 48 70

Class 12R: 35 52 46 66 36 38 44 65 38 69 57 61 43

23 61 25 44 50 51 60 57 69 58 44 72

a Create a five-number summary for this exam data and draw a box-and-whisker plot for each class.

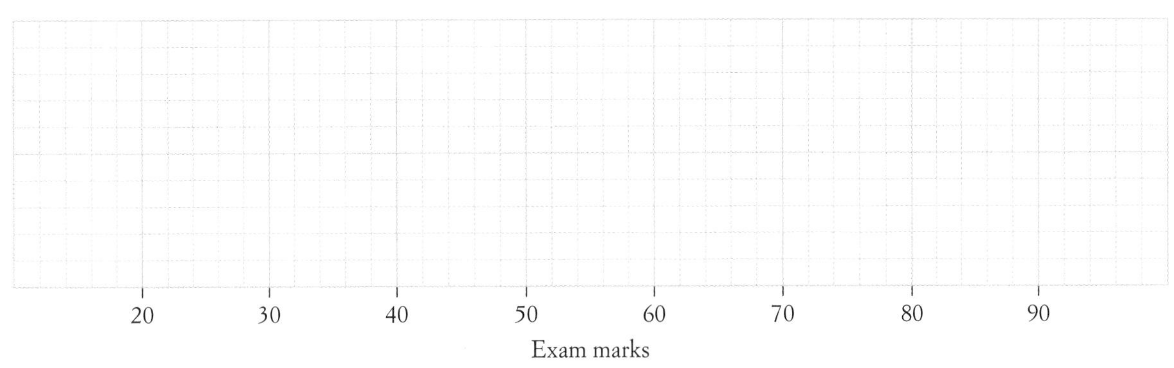

b Which class performed better in the exam? ______

c Which class had a greater spread of marks? ______

d Which class had an outlier? What was the outlier mark? ______

e Describe the shape of the distribution of marks for each class.

______

f Evaluate the following claim: 'Although one class generally performed better than the other, this is not true if only the bottom 25% of marks for each class is compared.'

______

______

Optional: Use graphing software to construct box-and-whisker plots for each class' marks.

DATA CROSSWORD

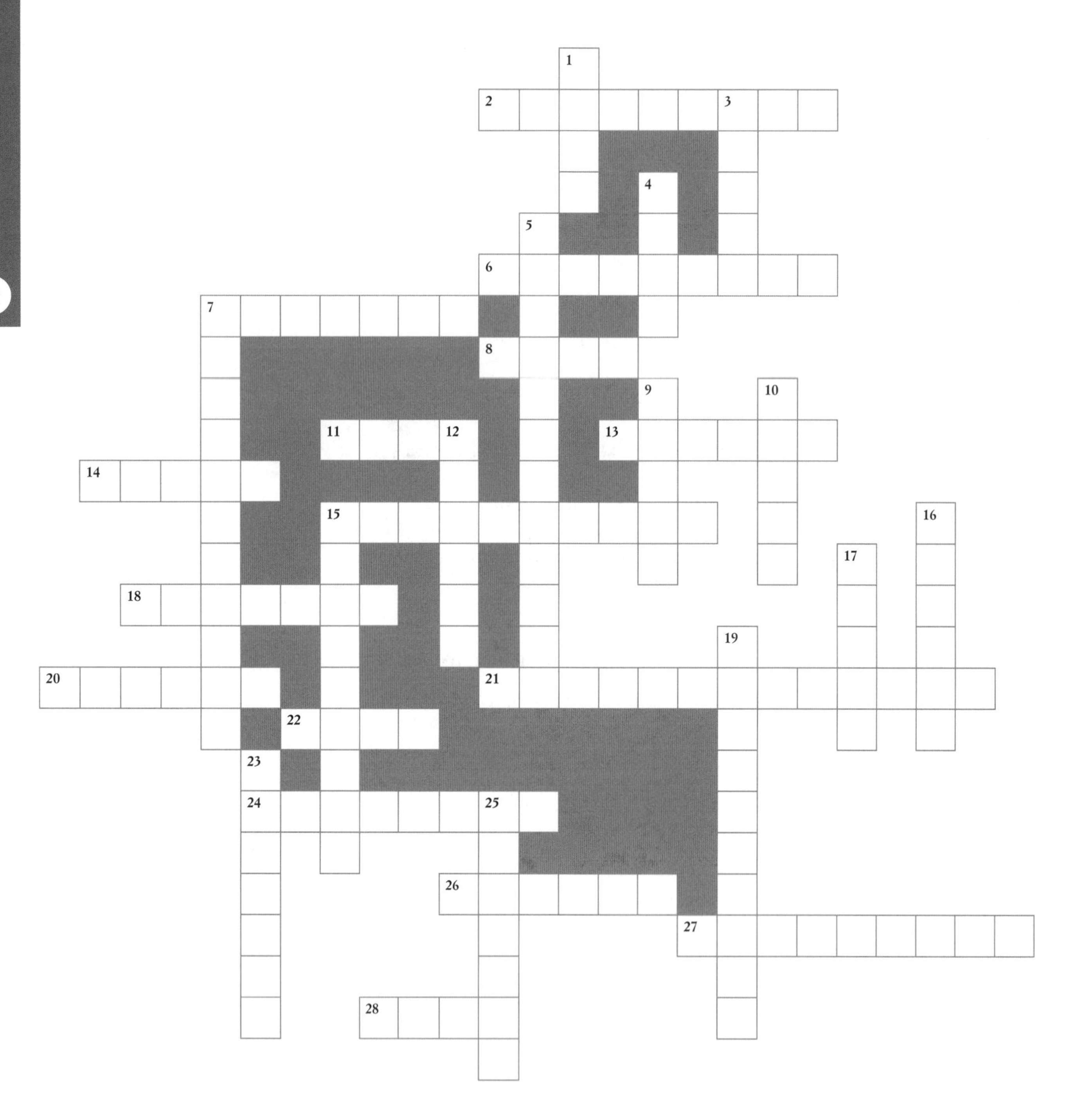

PS PUZZLE SHEET

Clues across

2 Q_1, Q_2, Q_3 are called these

6 Frequency column graph

7 A graph of points on a number plane is called a ________ plot.

8 ________-and-leaf plot

11 Something misleading that causes a sample or graph to not truly represent a situation

13 Interquartile range is a measure of ________

14 To find the median, first sort the scores in ________

15 A boxplot illustrates a ________-________ summary

18 Box-and-________ plot

20 The average of the 2 middle values if there is an even number of values in a data set

21 This range is the difference between Q_3 and Q_1

22 Sum of the data values divided by the number of values

24 Mean, median, mode are measures of ________ (or central tendency).

26 A scatterplot of points that approximate a line indicates a ________ relationship between 2 variables

27 Data in the form of ordered pairs that measure 2 variables

28 Most frequent value in a data set

Clues down

1 Another word for information

3 Q_1 is called the ________ quartile

4 A simple graph is the dot ________

5 'D' word meaning set of values

7 Not skewed

9 The highest value is also called the ________ extreme

10 The simplest measure of spread

12 A 'twisted' distribution of data values

15 How often a value appears in a data set

16 The median represents the ________ value

17 The interquartile range is the range of the middle ________ %

19 'cf' stands for ________ frequency

23 Describes where many of the values in a data set are grouped

25 An extreme data value that is much different to the others

5 DATA

Name:

Due date:

Parent's signature:

Part A	/ 8 marks
Part B	/ 8 marks
Part C	/ 8 marks
Part D	/ 8 marks
Total	/ 32 marks

REMEMBER THAT MEDIAN MEANS 'MIDDLE VALUE' WHILE MODE IS 'MOST OFTEN'.

PART A: MENTAL MATHS

Calculators not allowed

1 Which fraction is smaller: $\frac{7}{10}$ or $\frac{7}{12}$?

2 Simplify each expression.

a $\frac{3n}{10} - \frac{2n}{5}$

b $\frac{p}{3} \div \frac{3p}{7}$

3 A truck is travelling at an average speed of 95 km/h. How far will it travel in 5 hours?

4 (4 marks) Find the area of each shape:

a as a surd

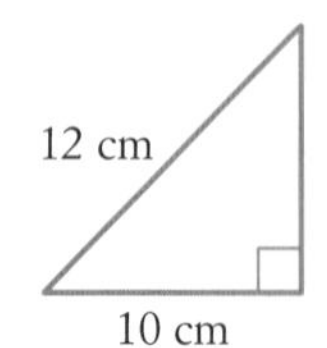

b in terms of π

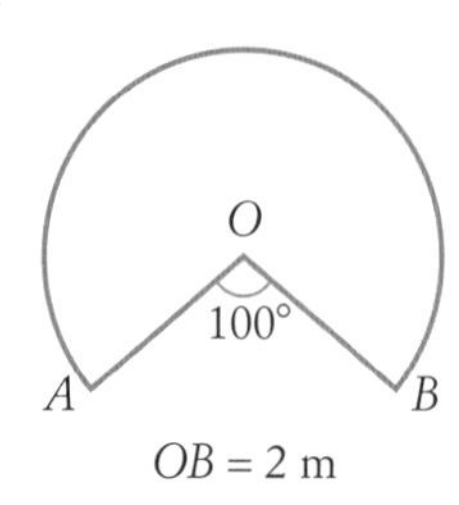

PART B: REVIEW

1 For this set of data, find:

Score	Frequency
1	11
2	7
3	12
4	8
5	12
Total	

a the range

b the mean

c the median

d the mode

2 For this set of data, find:

53 40 26 33 30 38 27 35 39 28

a the median

b the mean

c the range

d the outlier

HW HOMEWORK

C S F

9780170454551

PART C: PRACTICE

› The shape of a distribution
› Interquartile range

1 Describe the shape of each distribution.

a

0 1 2 3 4

____

b

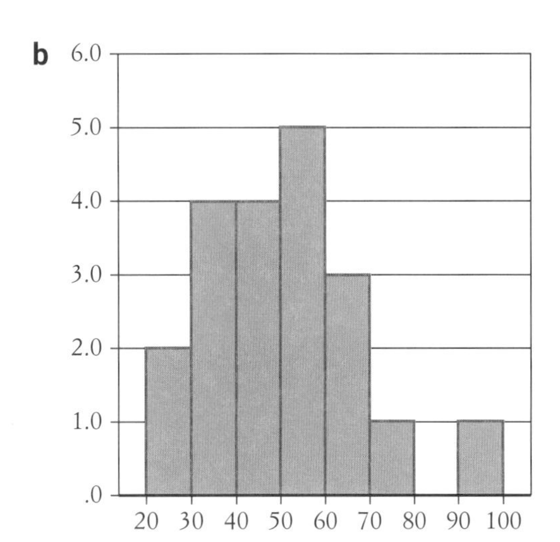

____

c

Stem	Leaf
0	8
1	2 5
2	2 4 8
3	0 3 3 6 8
4	0 5 5

____

2 For this data set, find:

8 10 12 14 16 16 17
17 19 21 24 25 28

a the lower quartile ______________

b the median ______________

c the upper quartile ______________

d the interquartile range ______________

e the mode ______________

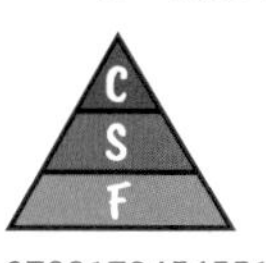

PART D: NUMERACY AND LITERACY

1 **a** The range and interquartile range are both measures of ______________.

b A distribution is negatively skewed if its tail points to the ______________.

c A ______________ distribution has 2 peaks.

d An ______________ is a data value that is much different from the other values in the data set.

e ______________ is the quartile that divides the lower 25% of scores.

2 For the data in this dot plot, find:

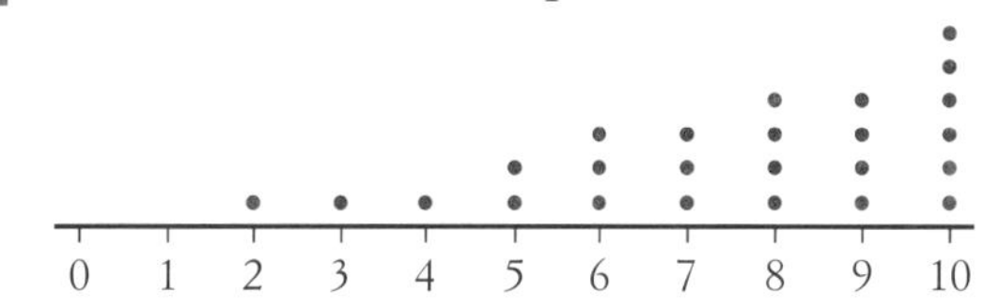

a the shape of the distribution

____

b the median

____

c the interquartile range

____

HW HOMEWORK

9780170454551

5 BOXPLOTS

Name:

Due date:

Parent's signature:

Part A	/ 8 marks
Part B	/ 8 marks
Part C	/ 8 marks
Part D	/ 8 marks
Total	/ 32 marks

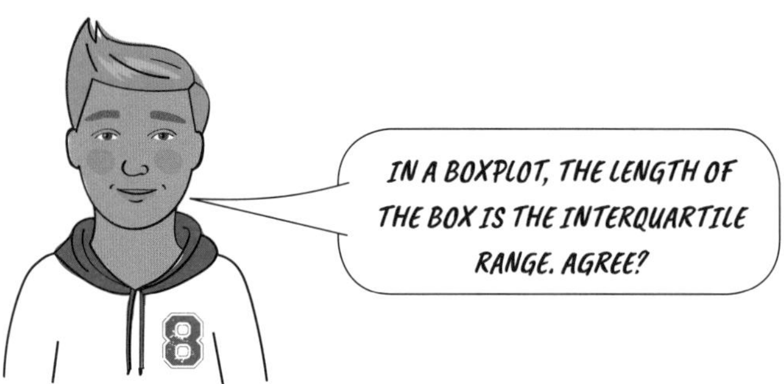

HW HOMEWORK

PART A: MENTAL MATHS

Calculators not allowed

1 Evaluate each expression if $a = -3$, $b = 2$ and $c = 5$.

a $\sqrt{a+b+c}$ __________

b $\dfrac{a}{c-b}$ __________

2 (2 marks) Simplify $\dfrac{6wy}{zx} \div \dfrac{10y}{xz} \div \dfrac{5z}{w}$.

____

3 For $\triangle ABC$, find the value of:

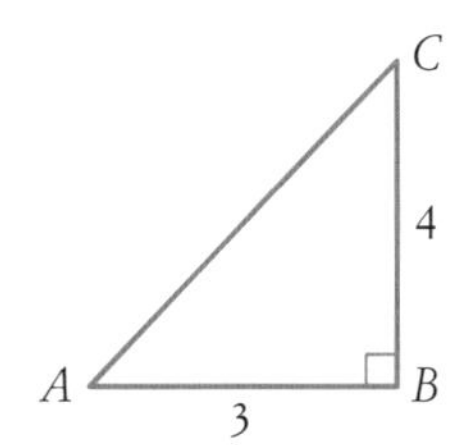

a AC

____

b $\cos C$ __________

4 (2 marks) Find x, giving reasons.

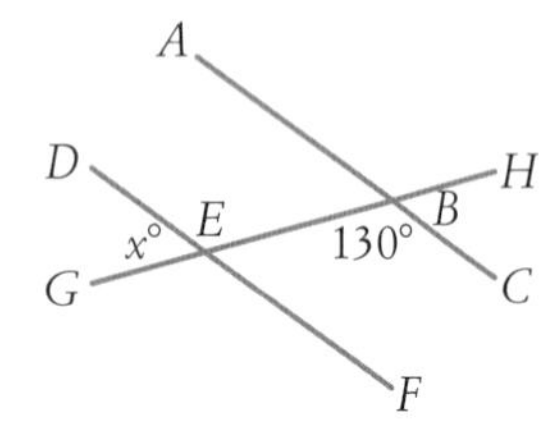

____

____

PART B: REVIEW

For this data set, find:

Stem	Leaf
1	0 3 6
2	1 6 7 8
3	5 5 6
4	1 1 5 6 9
5	0 3 6 8

1 the range __________

2 the median __________

3 the mean (to 2 decimal places)

____

4 the lowest value __________

5 the lower quartile (Q_1) __________

6 the upper quartile (Q_3) __________

7 the highest value __________

8 the interquartile range __________

C S F

9780170454551

PART C: PRACTICE

› Boxplots

1 For this data set, find:

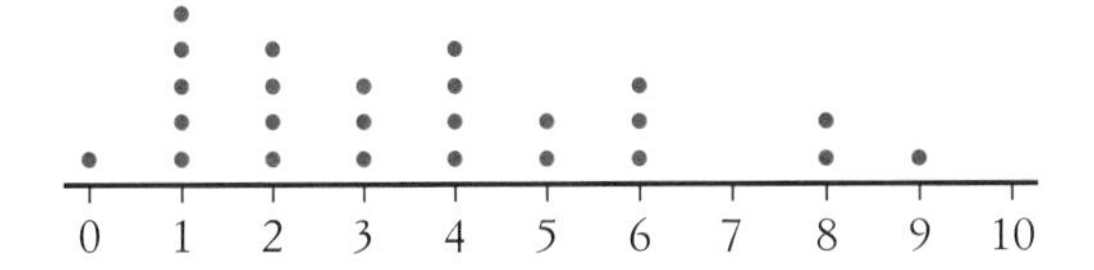

a the lowest value ____________

b the lower quartile ____________

c the median ____________

d the upper quartile ____________

e the highest value ____________

f the interquartile range ____________

2 (2 marks) Represent the data from question **1** on a boxplot.

PART D: NUMERACY AND LITERACY

1 (2 marks) Complete each box with a percentage.

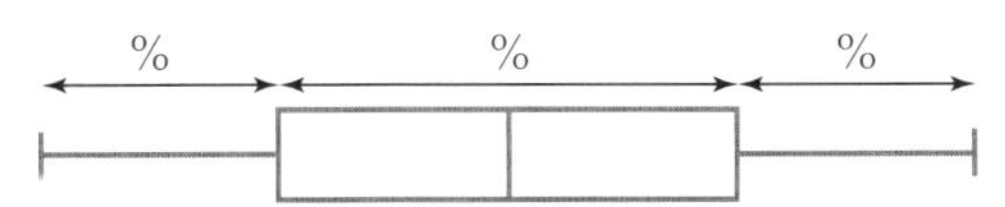

2 On a boxplot, what value is represented:

a on the end of the right whisker?

____

b on the left edge of the box?

____

c on the vertical bar inside the box?

____

3 (3 marks) Name the 5 things found in a five-number summary.

____

____

____

____

____

HW HOMEWORK

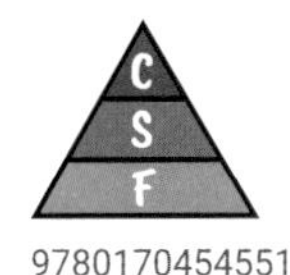

5 COMPARING DATA

Name:

Due date:

Parent's signature:

Part A	/ 8 marks
Part B	/ 8 marks
Part C	/ 8 marks
Part D	/ 8 marks
Total	/ 32 marks

PART A: MENTAL MATHS

Calculators not allowed

1 For this triangle, find the value of:

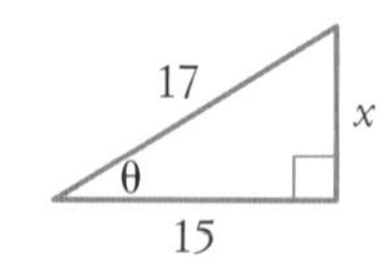

a x ______

b $\cos\theta$ ______

c $3\tan\theta$ ______

2 (2 marks) The price of a computer monitor was reduced by 20%. Find the original price of the monitor if its sale price was \$160.

______

______

3 Simplify:

a $(-3ab^2)^3$ ______

b $3a^{-1}b^0$ ______

______

4 Solve $3x + 4 = 5x - 9$.

______

______

______

PART B: REVIEW

1 For this boxplot, find:

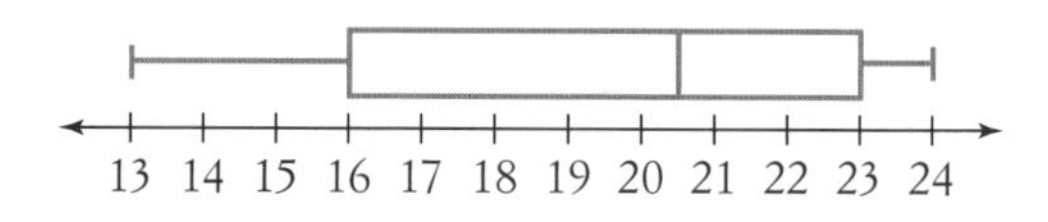

a the median ______

b the lowest value ______

c the upper quartile ______

d the interquartile range ______

2 (4 marks) Draw a boxplot for the data below, showing all values of the five-number summary.

9	2	12	9	12	5	4
2	8	8	1	6	2	10

HW HOMEWORK

9780170454551

PART C: PRACTICE

› Parallel boxplots
› Comparing data sets

1 (5 marks) These parallel boxplots show the heights (cm) of boys and girls in a Year 2 class.

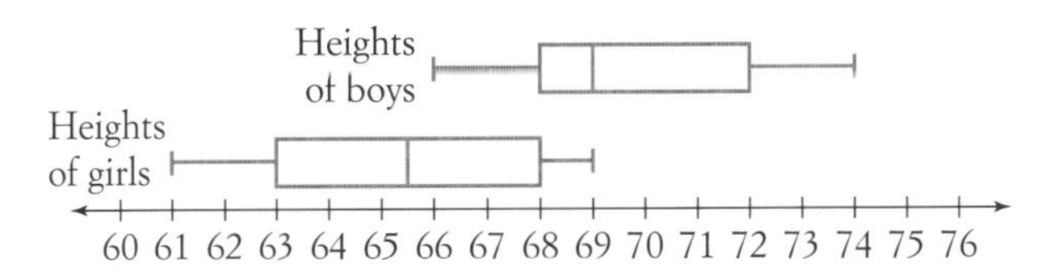

a Complete this table:

	Median	Interquartile range
Boys		
Girls		

b State one difference between the heights of boys and girls in this class.

2 The ages of males and females at a local library are shown in the stem-and-leaf plot.

Male		Female
	0	5 7 8
4 3 0	1	0 2 5 6 6 8
7 7 6	2	7
4 2	3	6
	10	3

a Find the mean for the males.

b Find the mean for the females.

c State one difference between the ages of males and females at the library.

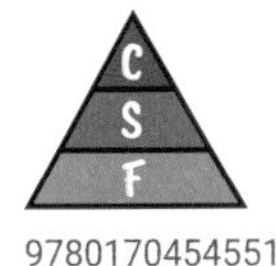

PART D: NUMERACY AND LITERACY

These back-to-back histograms show the ages of teachers at Ashfield and Burwood high schools.

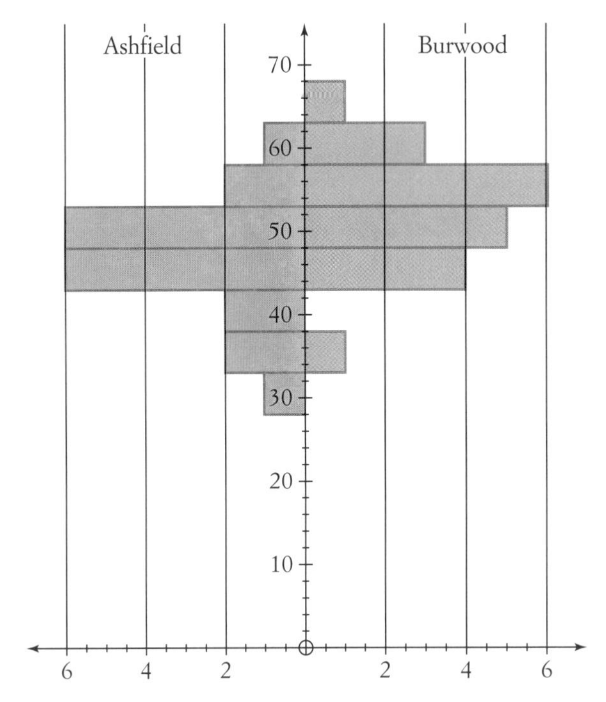

1 (5 marks) Complete this table:

Age group	Ashfield Frequency	Burwood Frequency
28– < 33	1	0
33– < 38		
38– < 43		
43– < 48		
48– < 53		
53– < 58		
58– < 63		
63– < 68		

2 Describe the shape of each set of data.

Ashfield:

Burwood:

3 State one difference between the ages of teachers at the 2 schools.

HW HOMEWORK

5 SCATTERPLOTS

SCATTERPLOTS ILLUSTRATE A PATTERN BETWEEN 2 QUANTITIES, SUCH AS AGE AND HEIGHT. HEIGHT IS THE DEPENDENT VARIABLE BECAUSE IT DEPENDS ON AGE.

Name:

Due date:

Parent's signature:

Part A	/ 8 marks
Part B	/ 8 marks
Part C	/ 8 marks
Part D	/ 8 marks
Total	/ 32 marks

PART A: MENTAL MATHS

Calculators not allowed

1 Write 46.2, 46.09 and 46.199 in ascending order.

____

2 Simplify the ratio 3 h : 30 min.

____

3 A block of land was bought for \$200 000, then sold for \$320 000 5 years later. Calculate:

a the profit ______________________

b the profit as a percentage of the cost price

____

4 (4 marks) Solve each equation.

a $-7x - 12 = -8 - 3x$

____

____

____

b $-10(2 - 5a) = 6(a - 3)$

____

____

____

PART B: REVIEW

The results of 3 topic tests are shown in the parallel boxplots. Name the test that has:

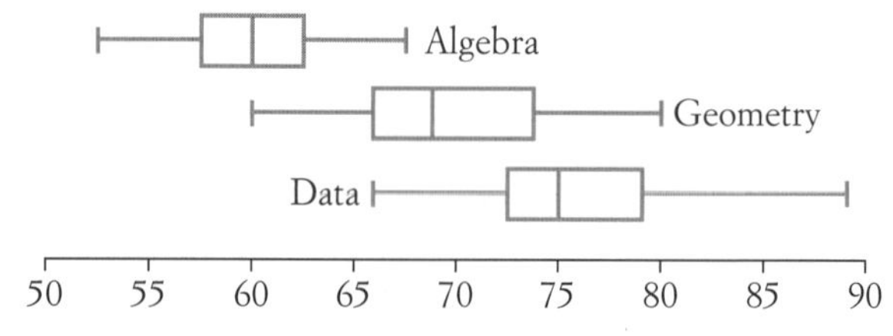

1 a range of 23 ______________________

2 a lowest score of 59 ______________________

3 a median of 69 ______________________

4 a symmetrical distribution ______________________

5 the most skewed distribution ______________________

6 the largest interquartile range ______________________

7 the best results ______________________

8 an interquartile range of 5 ______________________

HW HOMEWORK

C S F

9780170454551

PART C: PRACTICE

› Scatterplots

1 Describe the strength and direction of the relationship shown in each scatterplot.

a

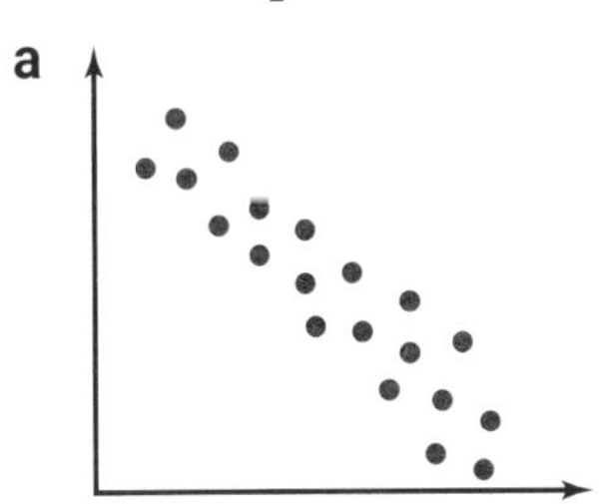

b

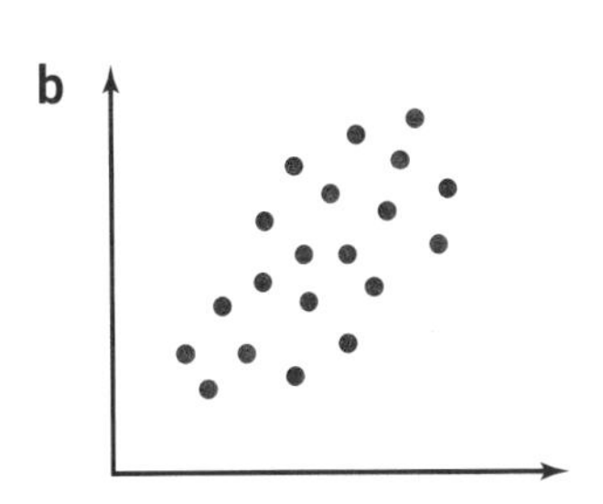

c

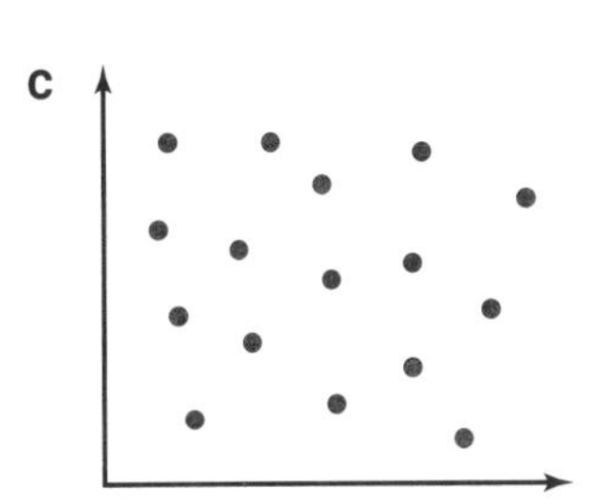

2 (5 marks) A survey was conducted on the age of people and their pulse (heartbeat) rate.

Age (A years)	45	20	15	62	35	49	18	38	55	24
Pulse rate (B beats/min)	71	79	83	70	68	67	86	72	68	86

a Graph the data as a scatterplot below.

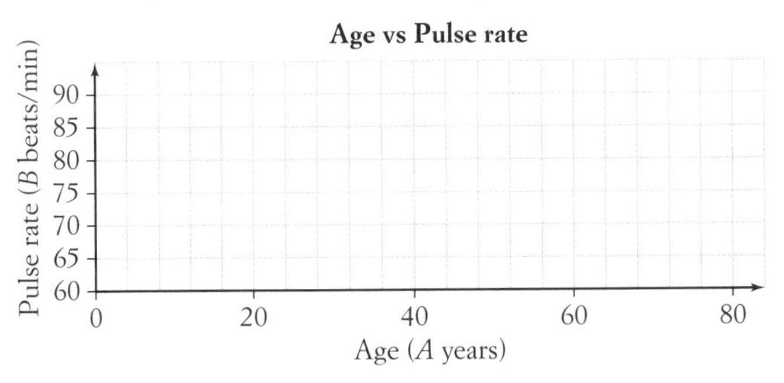

b What is the direction of the relationship between age of people and their pulse rate?

____

c Complete: As the age of the person increases, the pulse rate ______________

PART D: NUMERACY AND LITERACY

1 In the previous question, is pulse rate the dependent or independent variable?

____

2 What is data that measures 2 variables called?

____

3 What type of relationship exists between x and y if when x increases, y also increases?

____

4 Describe the strength of the relationship shown on a scatterplot if the points:

a are close together

____

b lie on a straight line

____

5 This dot plot shows the results of an online quiz (out of 10) for a group of students.

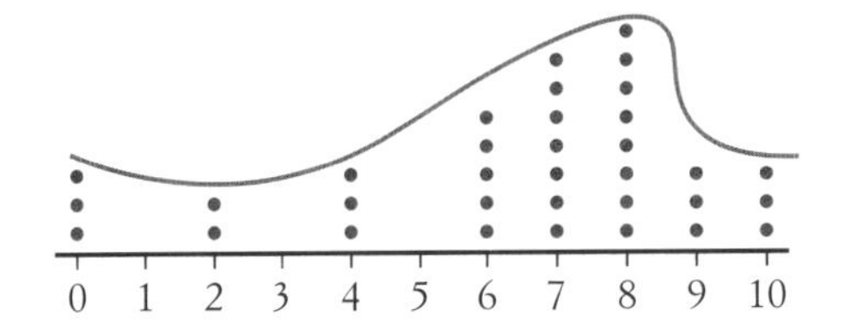

a Describe the shape of the data.

____

b What is the mode?

____

c How many students took the quiz?

____

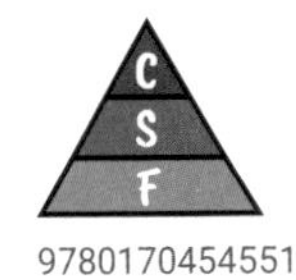

HW HOMEWORK

WS WORKSHEET

6 STARTUP ASSIGNMENT 6

PART A: BASIC SKILLS

/ 15 marks

1 Simplify 3^{-2}. ____________

2 Convert 75 km/h to m/s. ____________

____

3 Which is the better buy? ____________

A 250 mL for \$1.45

B 600 mL for \$3.60

4 Write the geometrical symbol for 'is congruent to'. ____________

5 Convert $\frac{5}{6}$ to a percentage. ____________

6 Find the value of p.

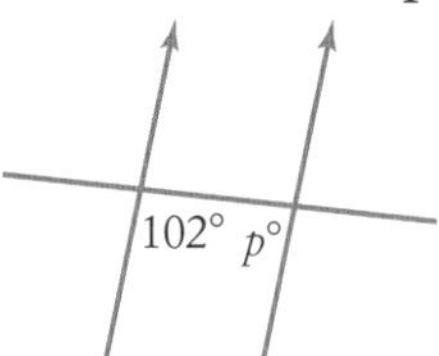

____

7 A map's scale is 1 : 50 000. What distance in kilometres does 4 cm on this map represent?

____

8 $A(-1, 2)$ and $B(4, 6)$ are points on the number plane. Find:

a the length of AB in surd form ____________

____

b the gradient of AB. ____________

9 Write 1 049 720 correct to 3 significant figures.

____

10 What is the most frequent value of a set of data called? ____________

11 Simplify 27 : 18 : 6.

____

12

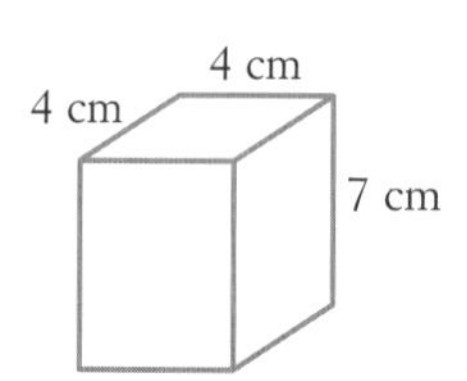

For this square prism, find its:

a surface area ____________

____

b volume. ____________

13 Write 0.006 in scientific notation. ____________

PART B: EQUATIONS

/ 25 marks

14 Solve $3p + 4 = 18$. ____________

15 $-1 > 6$, true or false? ____________

16 If $y = 3x - 4$, find y when $x = -1$. ____________

17 Expand:

a $2(5a - 3)$ ____________

b $-6(2a - 4)$ ____________

18 Test whether $x = 4$ is a solution to $5x + 6 = 3x + 14$. ____________

____

19 Solve $\frac{k}{4} = 4$. ____________

20 Solve $11 - 2r = 17$. ____________

____

C S F

9780170454551

21 Complete this table for $2x + y = 16$.

x	−2	−1	0	1	2	3
y						

22 Solve $\frac{2y}{3} + 6 = 10$. ______

______

23 Solve $-3m = -25$. ______

24 Expand and simplify:

a $2(r + 5) + 3(4r - 2)$ ______

______

b $(t + 9)(t - 1)$ ______

______

25 Solve $\frac{p + 7}{5} = 8$. ______

______

26 Solve $4(h - 2) = 30$. ______

______

27 Use an equation to find 3 consecutive odd numbers that add to 117.

______

______

______

______

28 Graph the line $y = 3x - 2$ on a number plane.

29 Find the size of the obtuse angle in this triangle.

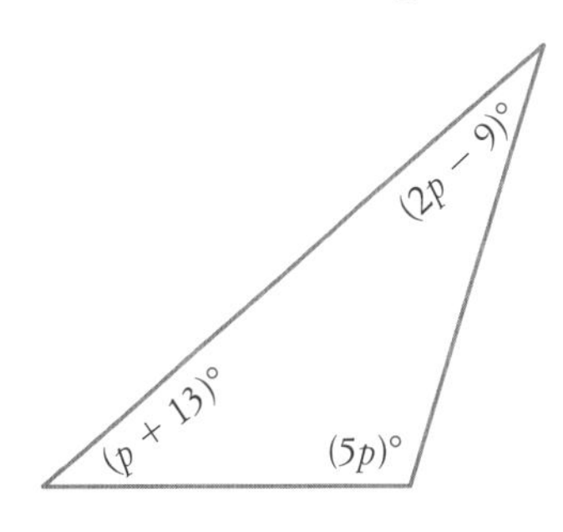

______

30 If $2x + 3y = 24$ and $y = 6$, find x. ______

______

______

31 The cost of hiring a karaoke machine is given by the formula $C = 90 + 25h$, where \$$C$ is the total cost and h is the number of hours.

a Find the cost of hiring the machine for 6 hours. ______

b How many hours will $165 pay for?

______

______

32 Test whether $n = -28$ is the solution to $\frac{n - 2}{5} - \frac{n}{4} = 1$. ______

______

______

33 Solve $3x - 5 = x + 9$. ______

______

______

PART C: CHALLENGE

Bonus / 3 marks

A farmer has emus and pigs. Altogether there are 250 creatures and they have 834 legs. How many of each animal is there?

______

C S F

9780170454551

6 GRAPHING INEQUALITIES

Graph each inequality on the number line provided.

1 $x \geq 3$

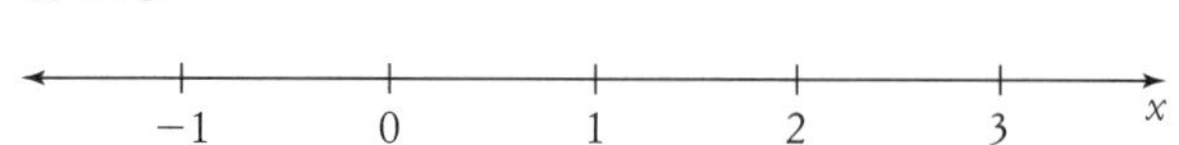

2 $x > 1$

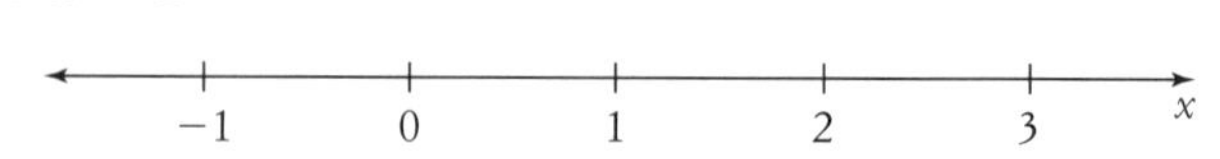

3 $x < -2$

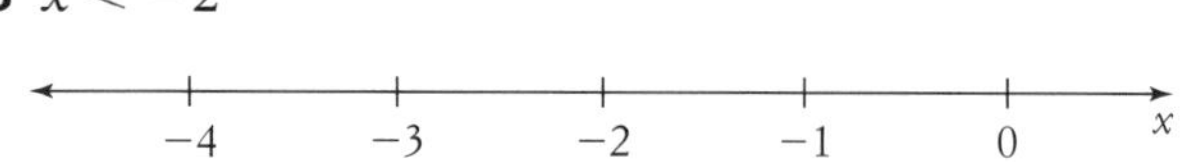

4 $x \leq 5$

5 $x > -1$

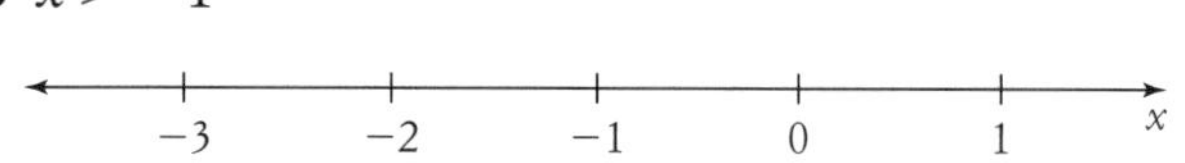

6 $x \geq 0$

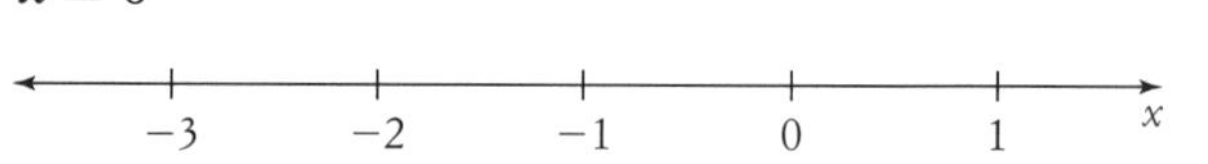

7 $x \leq -1$

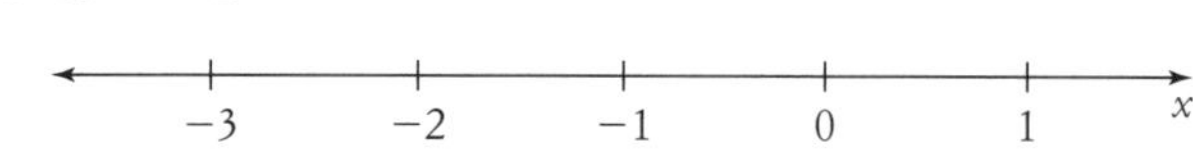

8 $x > 3\frac{1}{2}$

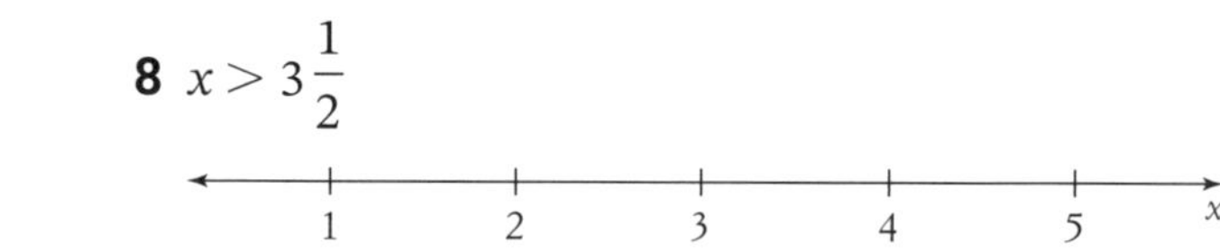

9 $x + 8 \leq 12$

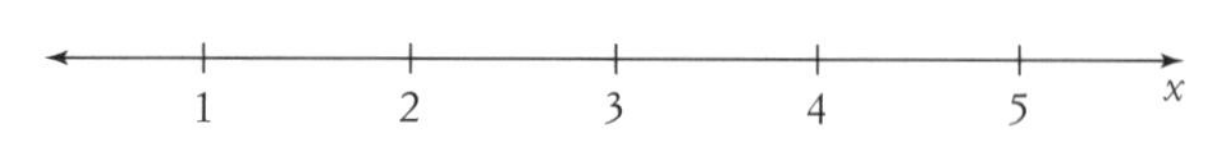

10 $4x + 5 \leq -1$

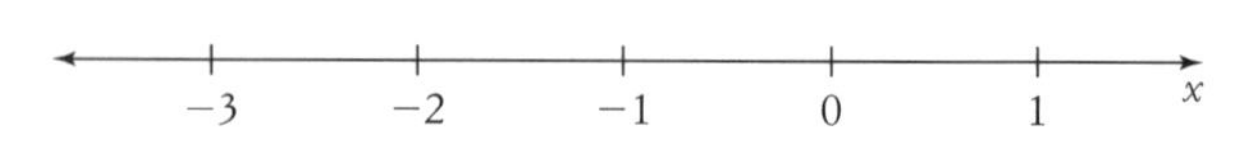

11 $7x - 2 < -2$

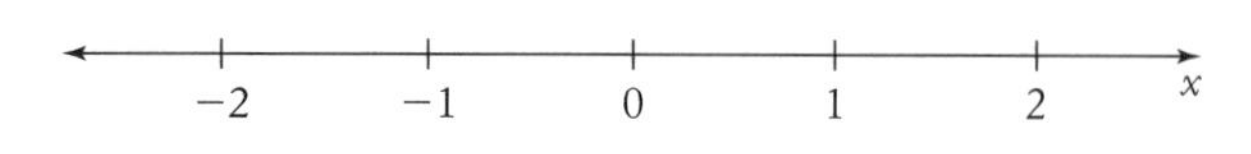

12 $8x \geq -24$

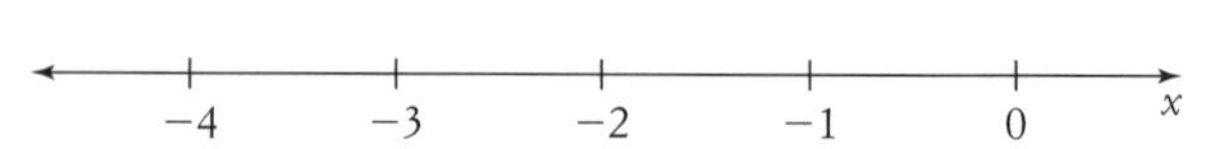

13 $x + 5 < 7$

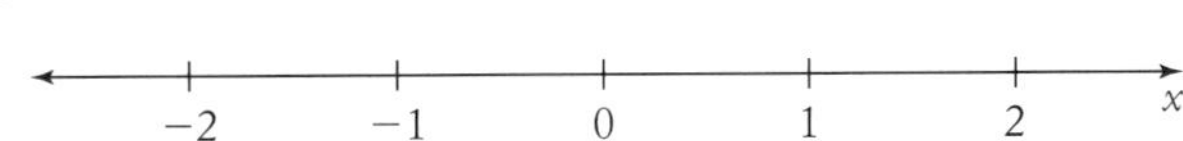

14 $3x > 15$

15 $8x - 1 \leq 11$

16 $2(x + 5) \geq -2$

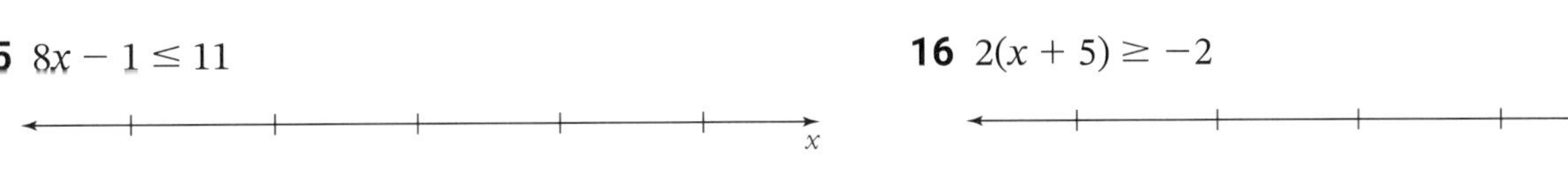

17 $4(x + 4) \geq 6$

18 $7(x - 4) < -7$

19 $\dfrac{x}{4} \leq 2$

20 $\dfrac{x + 10}{2} > 3$

21 $2(2x + 9) < 20$

22 $6(x - 7) \geq -9$

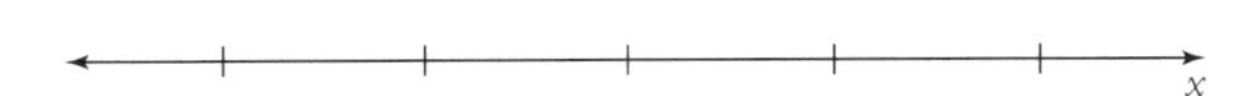

23 $\dfrac{x - 8}{6} \leq -2$

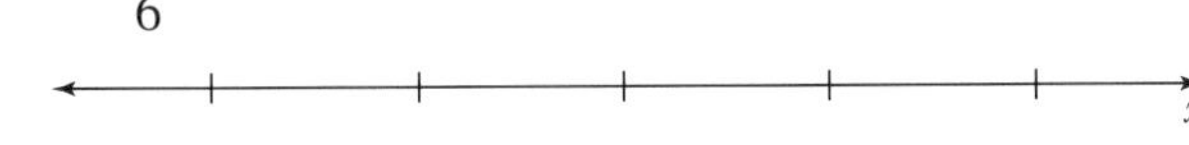

24 $\dfrac{2x + 7}{3} > 4$

6 EQUATIONS AND INEQUALITIES CROSSWORD

PS PUZZLE SHEET

9780170454551

Clues down

1 We must do this to $u^2 + 3u - 28$ before we solve $u^2 + 3u - 28 = 0$

2 $<$ means ______ than

5 [ ]

9 Inequalities can be graphed on a number ______.

10 The answer to an equation or inequality is called its ______.

11 Number less than 0

14 Most quadratic equations have ______ solutions

15 An equation involving x^2

18 Abbreviation for 'lowest common multiple'

21 A rule written as an equation

Clues across

3 The variable on the left side of a formula is called its s ______

4 To replace a variable with a number

6 To remove the brackets in an algebraic expression

7 $\geq$ means ______ than or equal to

8 $3x + 6 \leq 15$ is an example of one

12 To find the answer to an equation or inequality

13 $18 - 5x = 10$ is an example of one

16 $\sqrt{}$ means square ______

17 A letter that stands for a number

19 Equations can be solved by these 'opposite' operations

20 The M in LCM

21 It has a denominator

22 Abbreviation for 'right-hand side'

23 A simple way of solving equations is guess and ______

24 Abbreviation for 'left-hand side'

6 EQUATIONS

Name:

Due date:

Parent's signature:

Part A	/ 8 marks
Part B	/ 8 marks
Part C	/ 8 marks
Part D	/ 8 marks
Total	/ 32 marks

THE EQUATIONS ARE GETTING QUITE HARD NOW, BUT THAT'S BECAUSE THIS IS NEXT-LEVEL.

HW HOMEWORK

PART A: MENTAL MATHS

Calculators not allowed

1 Simplify each expression.

a $\frac{2}{3}\times\frac{9m}{10}$

b $\frac{8x}{3y}\div\frac{2x}{9y}$

c $8q^{-2}\times 7q$

d $9(t^{-1})^5$

2 **a** Increase \$140 by 5%.

b Decrease \$96 by 10%.

3 (2 marks) Find the value of x, giving reasons.

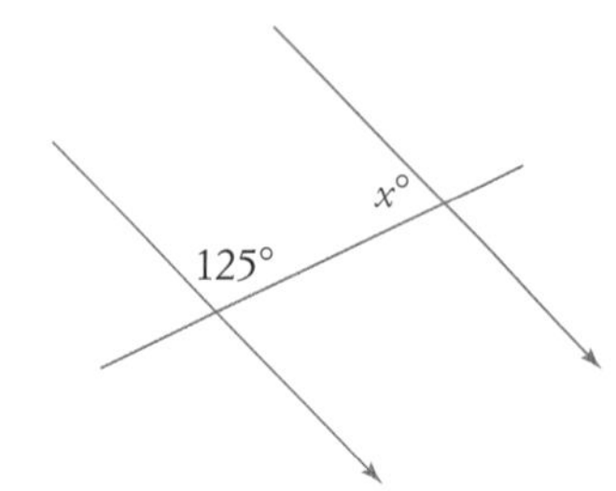

PART B: REVIEW

1 (6 marks) Solve each equation.

a $20 - 7y = 10 - 3y$

b $\frac{m}{3}-13=8$

c $9(b - 5) = 7(1 + b)$

2 (2 marks) When half of a number is decreased by 4, the result is 23. Use an equation to find the number.

C S F

9780170454551

PART C: PRACTICE

› Equations with algebraic fractions
› Quadratic equations $x^2 + bx + c = 0$

Solve each equation.

1 $\frac{3x+5}{6} = \frac{3}{4}$

2 $\frac{6m}{5} - \frac{2m}{3} = 4$

3 $a^2 - 18a + 77 = 0$

4 $y^2 - y - 20 = 0$

PART D: NUMERACY AND LITERACY

1 Complete: A quadratic equation is an equation in which the highest power of the variable is ______.

2 What is the inverse operation to squaring?

3 Explain why:

a $x^2 = 4$ has 2 solutions

b $x^2 = -4$ has no solutions

4 (4 marks) Solve each equation.

a $k^2 + 4k = 0$

b $\frac{4y-1}{5} = \frac{2y+4}{3}$

HW HOMEWORK

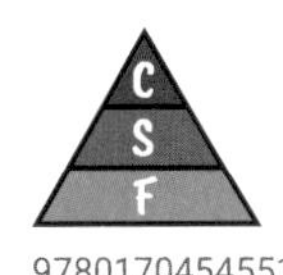

6 EQUATIONS AND FORMULAS

Name:

Due date:

Parent's signature:

Part A	/ 8 marks
Part B	/ 8 marks
Part C	/ 8 marks
Part D	/ 8 marks
Total	/ 32 marks

PART A: MENTAL MATHS

Calculators not allowed

1 Kathie bought a laptop for $800 and sold it at a profit of 8%.

a What was the profit?

b For what price did Kathie sell the laptop?

2 Factorise:

a $-20mn + 25n^2$

b $-x(3 + y) - 4(3 + y)$

3 (4 marks) Find the value of each variable, giving reasons.

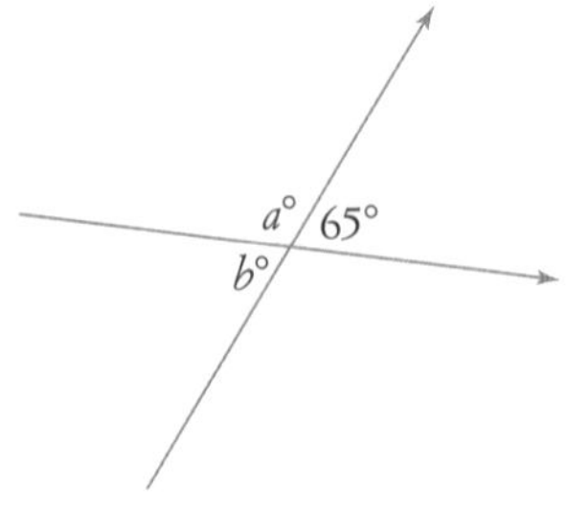

PART B: REVIEW

1 (4 marks) Solve each equation.

a $2a = 9(a - 5)$

b $5(3x + 2) = -40$

2 (4 marks)

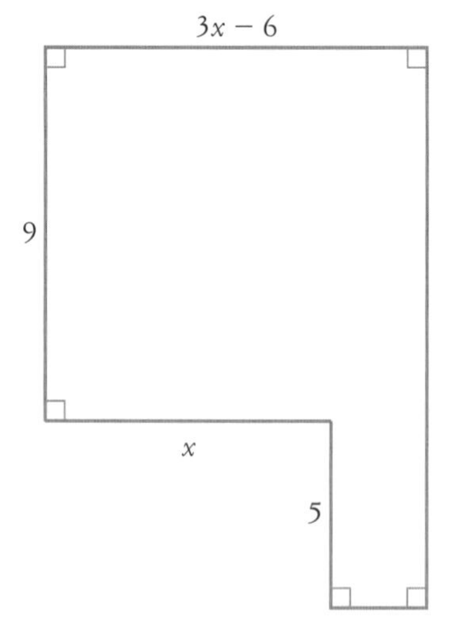

a Find x if the perimeter of this shape is 46 cm.

b Find the area of the shape.

C S F

9780170454551

PART C: PRACTICE

› Equation problems
› Equations and formulas

1 The formula $F = \frac{9C}{5} + 32$ converts a temperature in °C (Celsius) to °F (Fahrenheit). Convert 25°C to °F.

2 (3 marks) **a** Find x.

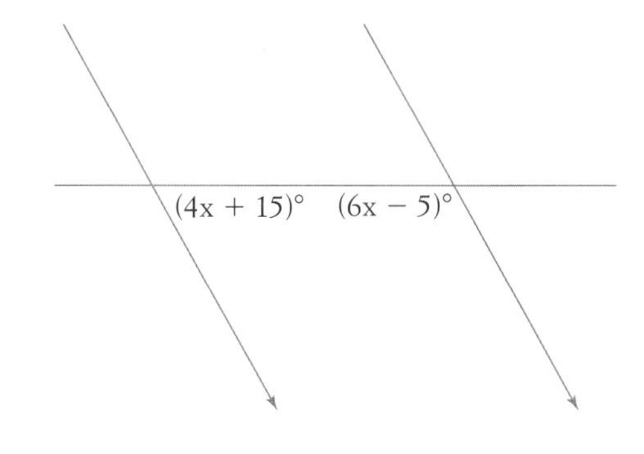

b Find the size of the acute angle.

3 (4 marks) The surface area of a closed cylinder has the formula $SA = 2\pi r^2 + 2\pi rh$. Calculate, correct to one decimal place:

a the surface area of a cylinder with radius 4.5 cm and height 6.2 cm

b the height of a cylinder with surface area 2104.23 cm^2 and radius 12.3 cm

PART D: NUMERACY AND LITERACY

1 (2 marks) If 8 more than a number is the same as 3 more than double the number, what is the number?

2 (3 marks) Gianni's bag contained only 10c and 50c coins. He had 124 coins, with a total of \$55.60. How many 10-cent coins did he have?

3 (3 marks) The sum of Sanjeev's and her mother's ages is 65. In 3 years, 3 times Sanjeev's age less 21 will be the same as her mother's age. How old is Sanjeev?

HW HOMEWORK

C
S
F

6 INEQUALITIES

SOLVING INEQUALITIES IS A NEW SKILL, BUT THE RULES ARE SIMILAR TO THOSE FOR SOLVING EQUATIONS.

Name:

Due date:

Parent's signature:

Part A	/ 8 marks
Part B	/ 8 marks
Part C	/ 8 marks
Part D	/ 8 marks
Total	/ 32 marks

PART A: MENTAL MATHS

Calculators not allowed

1 Complete: 1500 cm = __________ m

2 Evaluate $7^2 + 9^2$ __________

3 Write the recurring decimal 0.124124124... using dot notation.

____

4 For this set of data, find:

22 24 26 24 23 30 31 22

26 27 31 24 25 22 21 29

a the mode

____

b the median

____

5 Find the area of each shape.

a

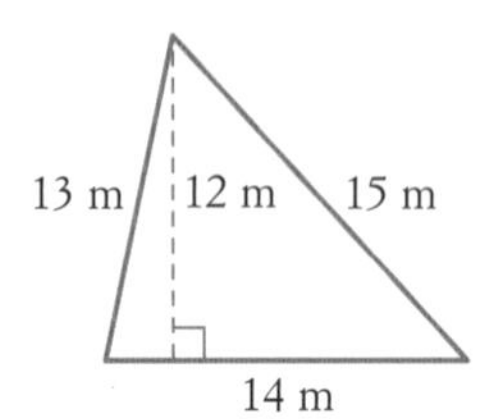

____

b

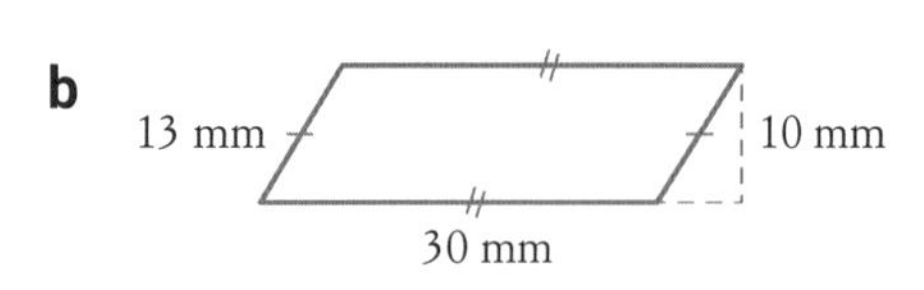

____

6 Write one property of the diagonals of a parallelogram.

____

____

PART B: REVIEW

1 (4 marks) Write each inequality in words and write a number that follows that inequality.

a $x > 4$

____

____

b $x \leq 4$

____

____

2 Graph each inequality on a number line.

a $x \leq -1$

____

b $x > 7$

____

c $x \geq 0$

____

d $x < 4$

____

9780170454551

PART C: PRACTICE

› Graphing inequalities on a number line
› Soliving inequalities

1 Write the inequality illustrated by each number line.

a

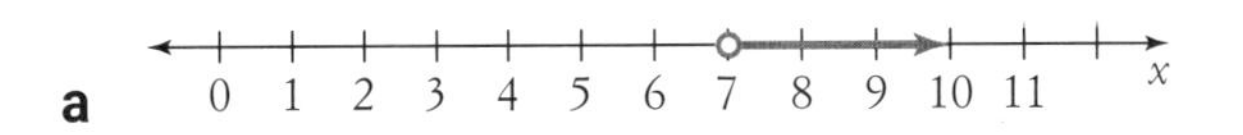

______

b

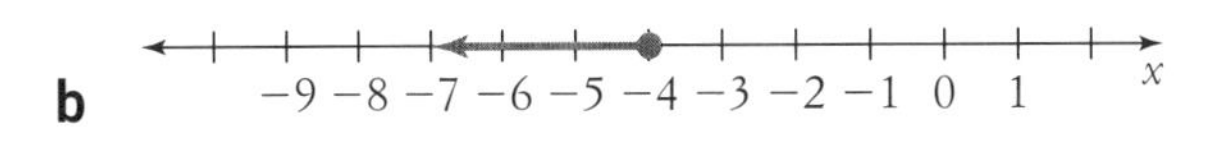

______

2 (6 marks) Solve each inequality.

a $3m - 11 \leq 10$

______

______

b $5(x - 5) > -30$

______

______

c $\frac{y+6}{7} \geq -4$

______

______

PART D: NUMERACY AND LITERACY

1 What word means the answer to an equation or inequality? ______

2 True or false? $-3 \geq -3$.

______

3 (2 marks) Complete: When solving inequalities, if you ______ both sides by a negative number, you must ______ the inequality sign.

4 (4 marks) Solve each inequality and graph its solution on a number line.

a $\frac{m}{6} \leq -3$

b $-5 - 20x > 5$

HW HOMEWORK

C S F

9780170454551

6 EQUATION REVIEW

SOME MIXED PROBLEMS HERE. IF YOU CAN SOLVE THESE, THEN YOU HAVE MASTERED EQUATIONS.

WS WORKSHEET

Solve each equation.

1 $5(2y + 3) = 25$

2 $-3(a + 7) = 7$

3 $-2(8 - x) = -6$

4 $\dfrac{k + 4}{3} = 9$

5 $\dfrac{4d - 7}{4} = -2$

6 $x^2 = 36$

7 $3n^2 + 5 = 32$

8 $5(m - 1) + 3(2m + 3) = 26$

9 $3(3a + 4) = 5a - 4$

10 $\dfrac{p - 9}{7} + 5 = 6$

11 $\dfrac{2b + 11}{6} - 4 = 7$

12 $\dfrac{c}{3} + \dfrac{2c}{5} = 3$

13 $\dfrac{3e}{5} + 2 = \dfrac{1}{2}$

14 $\dfrac{y^2}{4} = 0.25$

15 $\dfrac{g^2 - 5}{2} = 10$

16 $\dfrac{4r}{7} - 5 = 9$

17 $4(3m - 2) = 6(m + 3)$

18 $\dfrac{2b^2}{9} = 14\dfrac{2}{9}$

19 $10(n - 4) = 5n + 10$

20 $\dfrac{8w - 5}{4} = \dfrac{9w + 4}{3}$

Mixed answers: $\pm 5, 4\frac{1}{3}, 23, -\frac{1}{4}, \pm 8, -2\frac{7}{12}, \pm 6, 1, 2\frac{1}{2}, -9\frac{1}{3}, \pm 3, 4\frac{1}{11}, -4, 10, 27\frac{1}{2}, \pm 1, 24\frac{1}{2}, 16, 2, 5$

STARTUP ASSIGNMENT 7 (7)

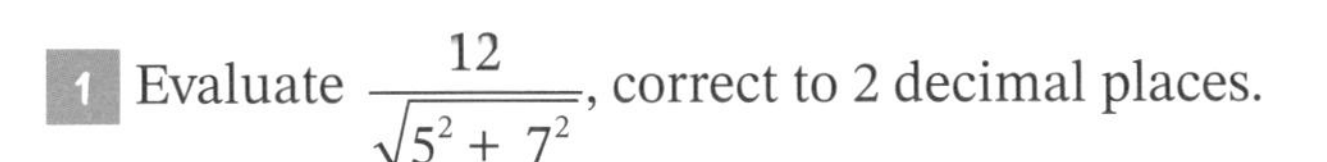

PART A: BASIC SKILLS

/ 15 marks

1 Evaluate $\frac{12}{\sqrt{5^2 + 7^2}}$, correct to 2 decimal places.

____

2 Simplify $4^{\frac{1}{2}} \times 4^{-2}$. ______________________

3 Complete: $4 \text{ m}^2 =$ __________ cm^2

4 Complete: 55 m/s = __________ km/h

5 Find the value of d in the diagram below, correct to 2 decimal places.

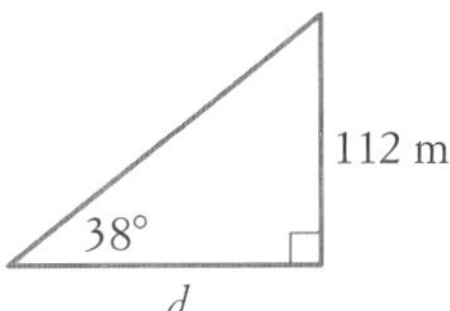

____

6 Simplify $\frac{20x^3y^2}{4x^3y}$. ______________________

7 Find, correct to 3 significant figures, the volume of a cylinder of radius 10 cm and height 12 cm.

____

8 Evaluate $\frac{2.46 \times 10^{12}}{4 \times 10^8}$. ______________________

9 What is the angle sum of a parallelogram?

____

10 If $\frac{5}{6}$ of a number is 90, what is the number?

____

____

11 A number is chosen at random from the numbers 1 to 20. What is the probability that it is a multiple of 3?

____

12 Find θ, to the nearest degree, if cos θ = 0.561.

____

13 Divide $4550 in the ratio 2 : 1 : 4. ______________________

____

____

14 How many axes of symmetry has a parallelogram? ______________________

15 Madeline is paid $275 plus a 5.5% commission on items sold. Calculate her pay when she sells $3300 worth of items.

____

PART B: ALGEBRA AND GRAPHS

/ 25 marks

16 **a** Complete the table below for $y = 3x - 2$.

x	−2	−1	0	1	2	3
y		−5			4	

b Graph $y = 3x - 2$ on a number plane.

c What is the gradient? ______________________

d What is the y-intercept? ______________________

e What is the x-intercept? ______________________

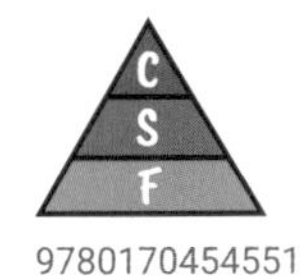

17 Complete the table below for $y = x^2 - 4$.

x	−2	−1	0	1	2	3
y		−3			0	

18 Solve each equation.

a $3x + 7 = 1$ ______________________

____

b $\frac{2y+4}{5} = 8$ ______________________

____

19 Draw a line with a positive gradient.

20 Evaluate $3x^2$ if:

a $x = 2$ ______________________

b $x = -4$ ______________________

c $x = -1$ ______________________

21 Write the formula for each table of values.

a

x	−2	−1	0	1	2	3
y	−5	−1	3	7	11	15

____

b

x	−2	−1	0	1	2	3
y	4	1	0	1	4	9

____

22 This graph shows Debbie's day trip.

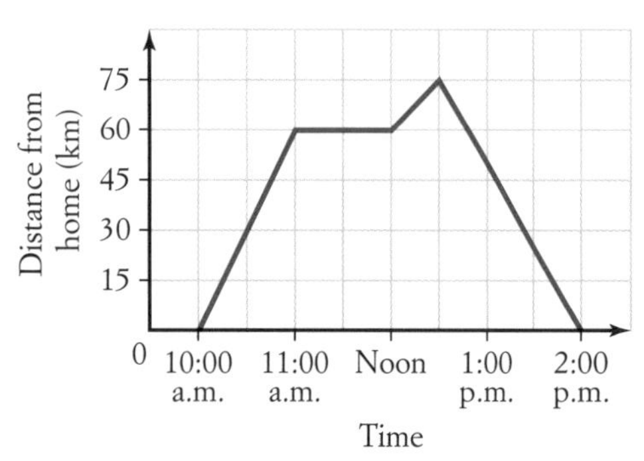

a What was Debbie's speed during the first hour? ______________________

b When did she begin her trip home?

____

c What total distance did she travel?

____

23 For this line, find:

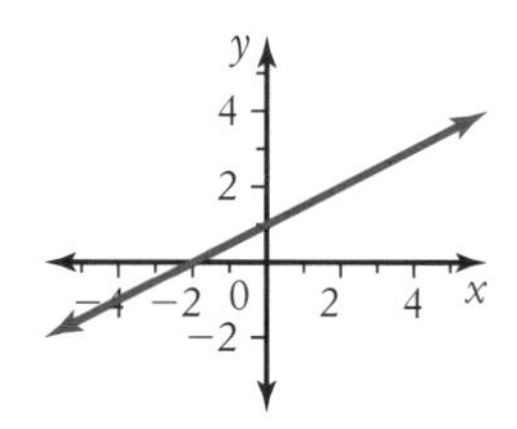

a the y-intercept ______________________

b the gradient ______________________

c its equation. ______________________

24 Evaluate $\frac{4}{x}$ if:

a $x = -3$ ______________________

b $x = 10$ ______________________

c $x = \frac{1}{2}$ ______________________

PART C: CHALLENGE

Bonus / 3 marks

A lady-bug walks along the edges of this prism, starting and ending its journey at vertex A.
What is the maximum distance it can walk without travelling along the same edge twice?

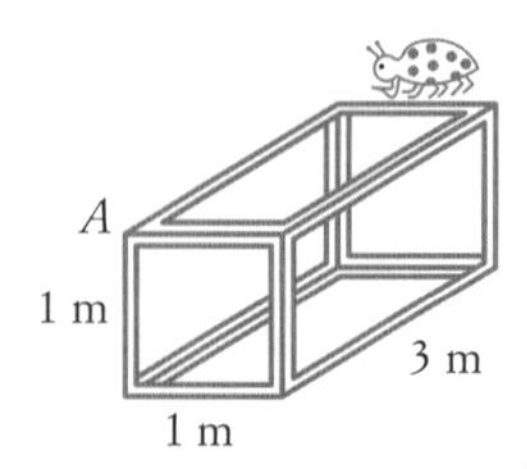

____

C S F

9780170454551

GRAPHING PARABOLAS 7

Teacher's tickbox

Graph the indicated set of equations.

- ❑ $y = x^2, y = -x^2, y = 2x^2, y = \frac{2}{3}x^2$
- ❑ $y = -x^2, y = -x^2 + 4, y = -x^2 - 3, y = -\frac{2}{3}x^2$
- ❑ $y = x^2, y = x^2 + 2, y = x^2 - 7, y = -2x^2$
- ❑ $y = x^2, y = 2x^2 + 1, y = 2x^2 - 3, y = -2x^2 + 3$

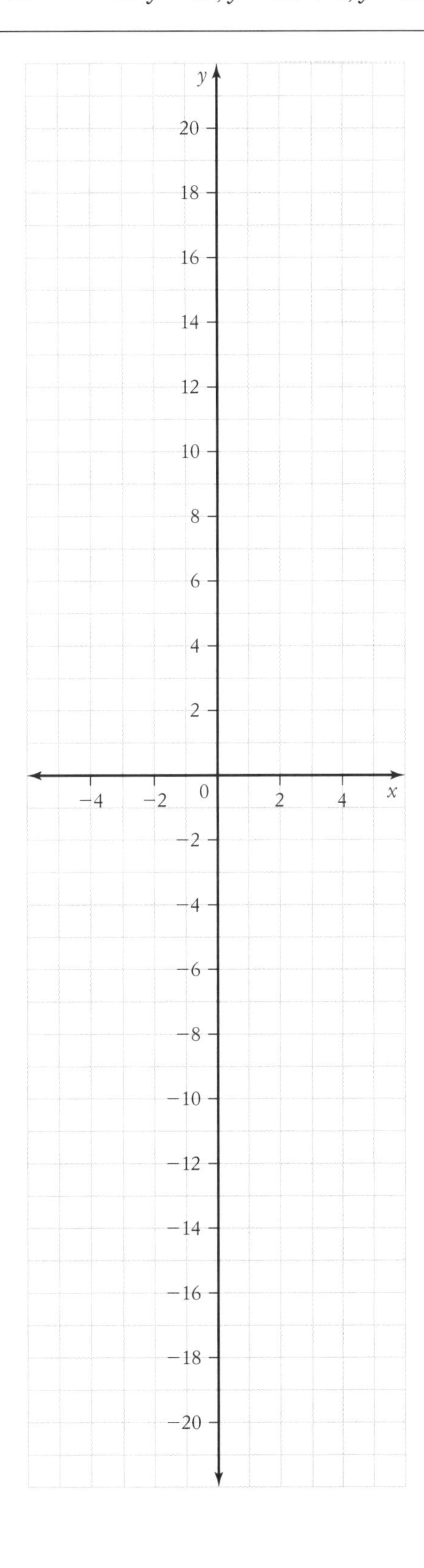

Teacher's tickbox

Graph the ticked set of exponential equations.

❑ $y = 2^x, y = 2^{-x}, y = -2^x, y = -2^{-x}$

❑ $y = 3^x, y = 3^{-x}, y = -3^x, y = -3^{-x}$

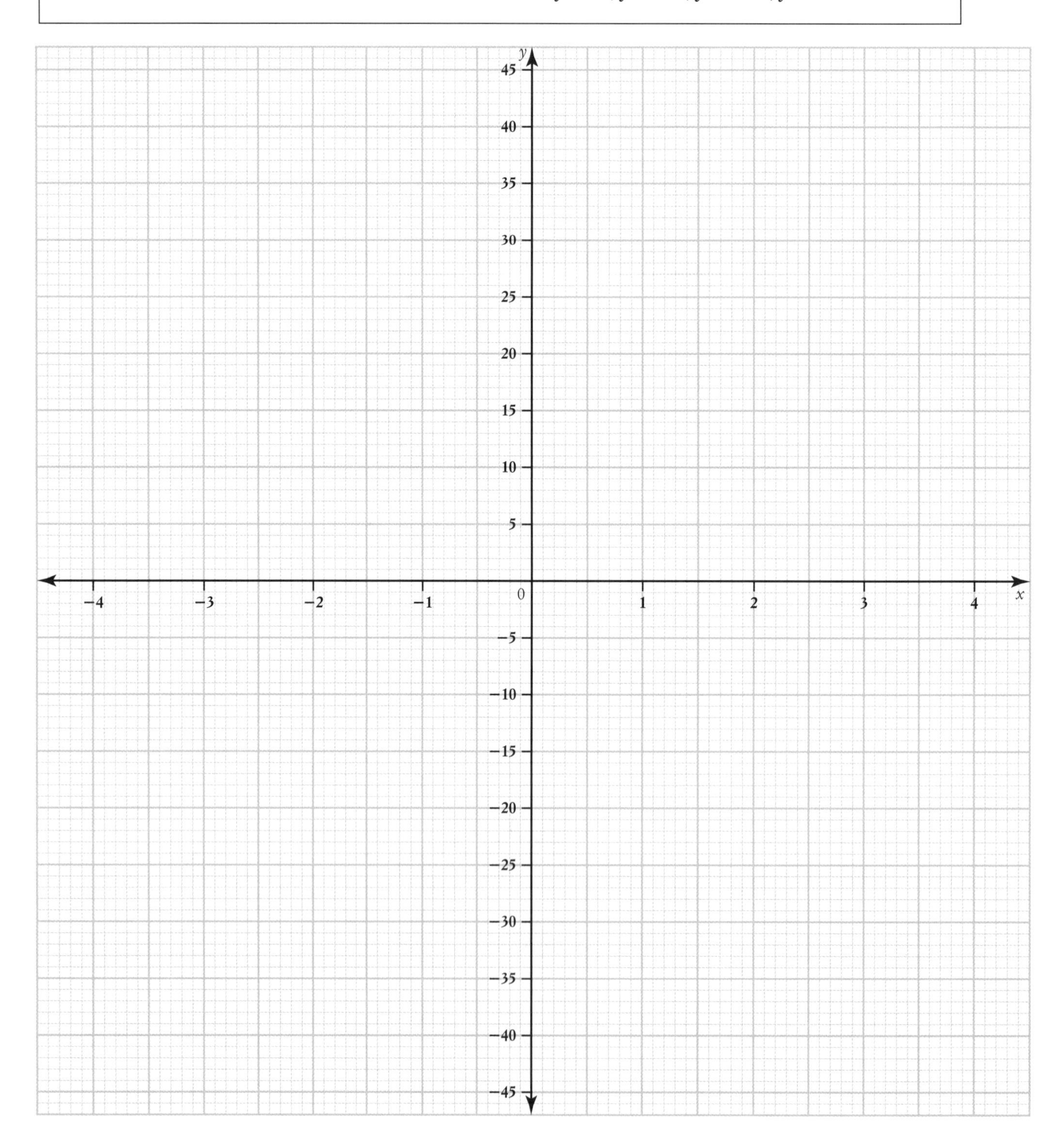

GRAPHING HYPERBOLAS 7

Teacher's tickbox

Graph the ticked pair of equations.

❑ $y=\frac{2}{x}, y=-\frac{1}{x}$ ❑ $y=\frac{2}{x}+1, y=\frac{3}{x}-2$ ❑ $y=\frac{3}{x}, y=-\frac{5}{x}$ ❑ $y=\frac{3}{x-1}, y=-\frac{2}{x+2}$

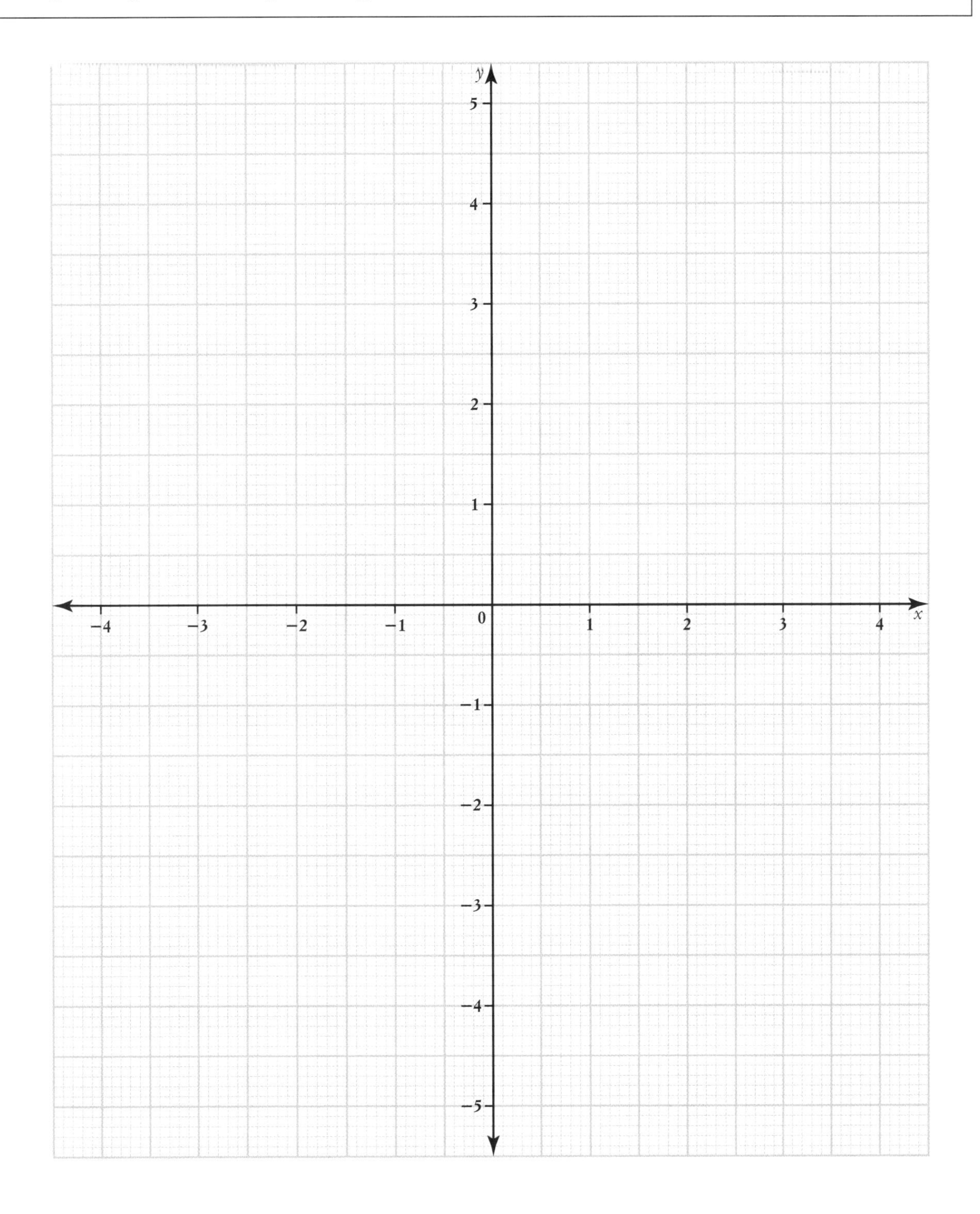

Unscramble the word clues and place them in the crossword.

9780170454551

Clues across

3 NEIL
6 TAMYESOPT
7 PHRAG
9 PINOT
10 CLERIC
11 RECENT
13 CREDIT
18 DECORATIONS
19 VEXTER
22 CANSNOTT
24 VERBIALA
25 HAZITLONOR
26 SIXA

Clues down

1 AUTOENQI
2 EVOCCAN
4 NIXANTELOPE
5 CARATQUID
8 CENTIOFFICE
11 VURCE
12 SNORNOVICE
14 NEVERIS
15 TALCRIVE
16 RIGION
17 PROIPROTON
20 SURADI
21 CENTREPIT
23 ALBOARPA

7 GRAPHING CURVES

Name:

Due date:

Parent's signature:

Part A	/ 8 marks
Part B	/ 8 marks
Part C	/ 8 marks
Part D	/ 8 marks
Total	/ 32 marks

PART A: MENTAL MATHS

Calculators not allowed

1 Convert $\frac{25}{40}$ to a percentage.

______

2 Simplify each expression.

a $(x + 3)(x - 4)$

______

b $\frac{a^3b^2}{3ab}$ ______

c $(6m)^{-2}$ ______

3 Name the 2 quadrilaterals that have all sides equal.

______

______

4 A jar contains only blue and white marbles. The probability of selecting a blue marble from the jar is 20%.

a What is the probability of selecting a white marble? ______

b If there are 25 marbles in the bag, how many blue marbles are there?

______

PART B: REVIEW

1 **a** If R is directly proportional to N and 0.74 is the constant of proportionality, write an equation for R.

______

b Find the value of R when $N = 5$.

______

c If this equation was graphed on a number plane, what type of graph would it be and what would be its y-intercept?

______

______

2 (2 marks) T varies directly with s. When $s = 5$, $T = 203.5$. Find T when $s = 24.1$.

______

______

______

3 (3 marks) Find the value of y when $x = 0$ in each equation.

a $y = 5x^2 + 8$

______

b $y = 5^x$

______

C S F

9780170454551

c $x^2 + y^2 = 16$

PART C: PRACTICE

› The parabola $y = ax^2 + c$
› The exponential curve $y = a^x$
› The circle $x^2 + y^2 = r^2$

1 (3 marks) Match each quadratic equation to its graph.

a $y = x^2$ **b** $y = -x^2 + 3$ **c** $y = x^2 - 1$

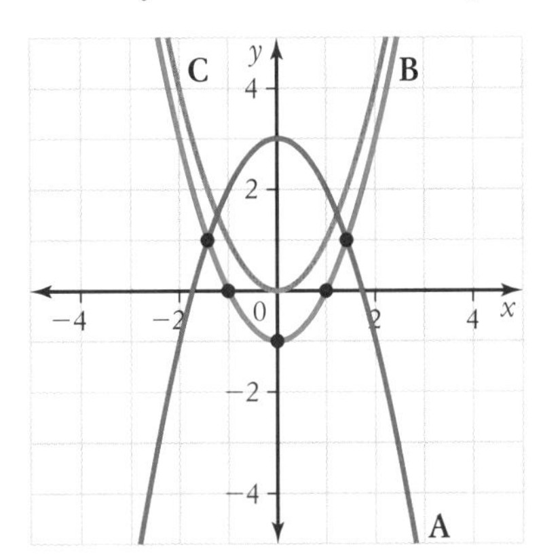

2 (4 marks) Match each exponential equation to its graph.

a $y = 2^x$ **b** $y = 2^{-x}$

c $y = -2^x$ **d** $y = -2^{-x}$

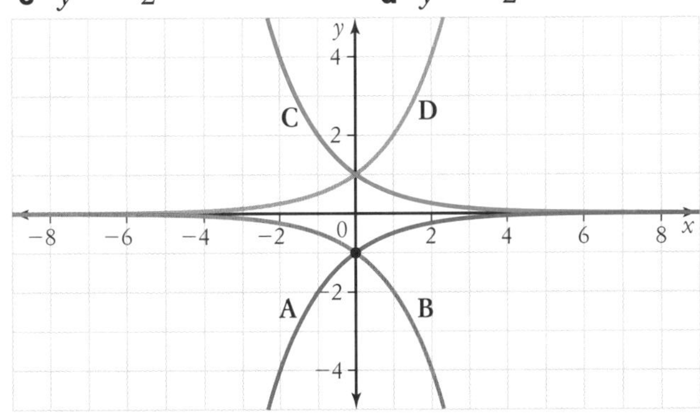

3 Write the equation of a circle with centre (0, 0) and radius 4.

PART D: NUMERACY AND LITERACY

1 **a** Is the graph of $y = -3x^2 + 1$ concave up or down? Give reasons.

b How many axes of symmetry does this graph have?

2 (2 marks) Graph $x^2 + y^2 = 9$ and state its centre and radius.

3 (2 marks) Sketch the graph of $y = 5^x - 1$ and state its asymptote.

4 (2 marks) Graph $y = \frac{1}{2}x^2 + 1$ and write the coordinates of its vertex.

HW HOMEWORK

C S F

WS WORKSHEET

8 STARTUP ASSIGNMENT 8

HI, I'M MS LEE. THIS ASSIGNMENT REVISES YOUR GEOMETRY SKILLS TO PREPARE YOU FOR THE TRIGONOMETRY TOPIC.

PART A: BASIC SKILLS

/ 15 marks

1 Simplify $\sqrt{\frac{50}{98}}$. ______

2 Felt-tip pens are on sale:

A 3 for $6.95

B 5 for $11.95

Which is better value, **A** or **B**? ______

3 Solve $\frac{x}{4} - \frac{2x}{3} = 5$. ______

______

4 15% of what amount is $10.80? ______

5 Complete: 1 m^2 = ______ cm^2

6 Write $\frac{1}{8}$ as a percentage. ______

7 Find the gradient of the interval joining (4, −1) and (7, −4). ______

8 For this rectangle, find:

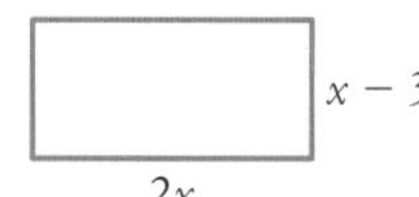

a the perimeter ______

b the area. ______

9 The angle sum of a polygon with n sides is $180(n - 2)°$. Find the size of one angle in a regular hexagon.

______

10 Simplify $m^6 \div m^2$. ______

11 Calculate the simple interest on a loan of $17 000 at 14% p.a. over 4 years. ______

12 Find the surface area of a cube with side length 5 cm. ______

13 Simplify 36 : 20. ______

14 Find the value of x.

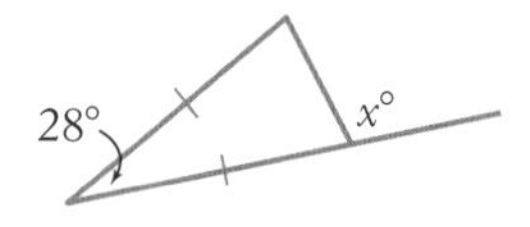

______

PART B: GEOMETRY

/ 25 marks

15 Round 18 min 42 s to the nearest minute.

______

16 How many degrees in 3 right angles?

______

17 Find the value of x in each diagram.

a

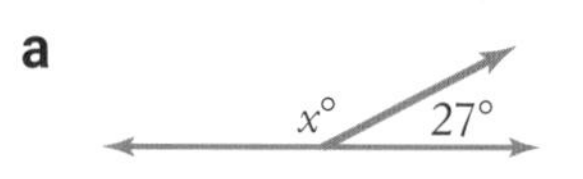

b

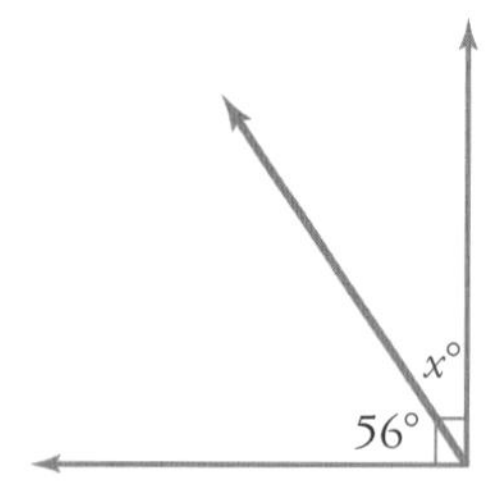

18 Evaluate:

a $4 \times \sqrt{24.01}$ ______

b $10 \times \sqrt{65.61}$ ______

c $\frac{63}{\sqrt{12.25}}$ ______

C S F

9780170454551

19 Name the hypotenuse in each triangle.

a

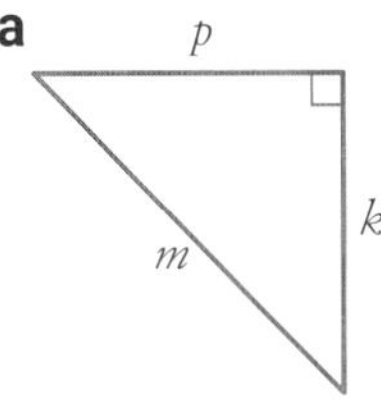

b

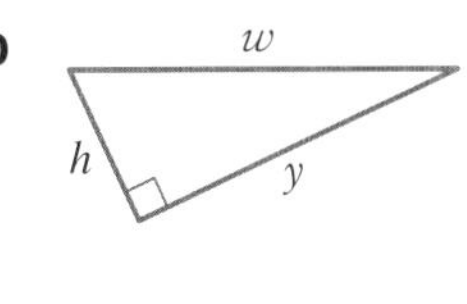

20 Convert 7.15 hours to hours and minutes.

21 **a** Find the value of x in each diagram.

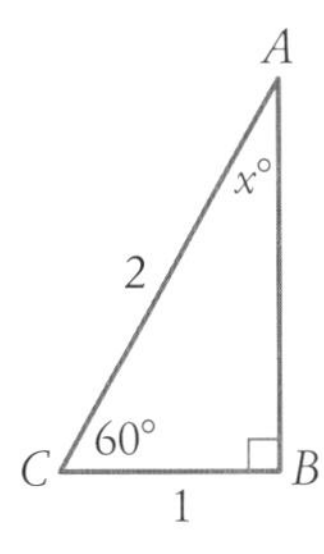

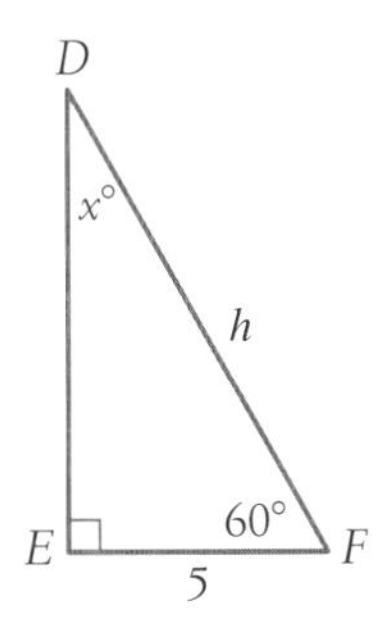

b If $\triangle ABC$ and $\triangle DEF$ above are similar, find h.

22 Find the value of r.

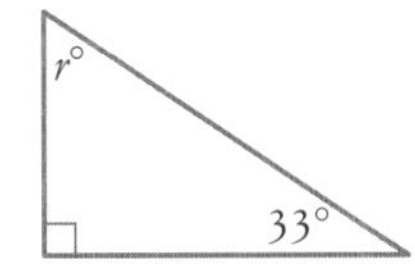

23 Find the value of b.

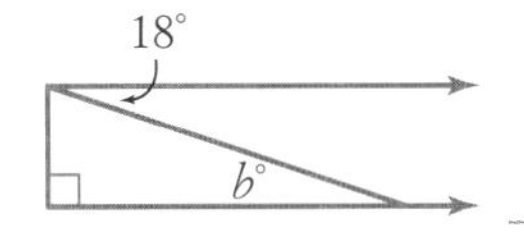

24 Solve each equation.

a $7 = \dfrac{x}{5}$

b $\dfrac{h}{2} = 10$

c $6 = \dfrac{30}{d}$

d $4 = \dfrac{10}{y}$

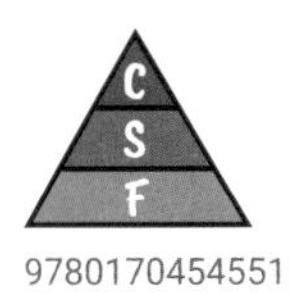

25 If $\angle Z = a°$, write an expression for the size of $\angle Y$.

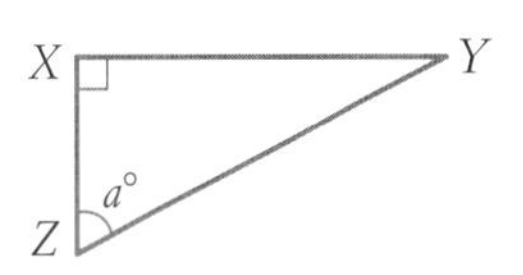

26 If $\triangle DEF$ is equilateral, find the values of x and y.

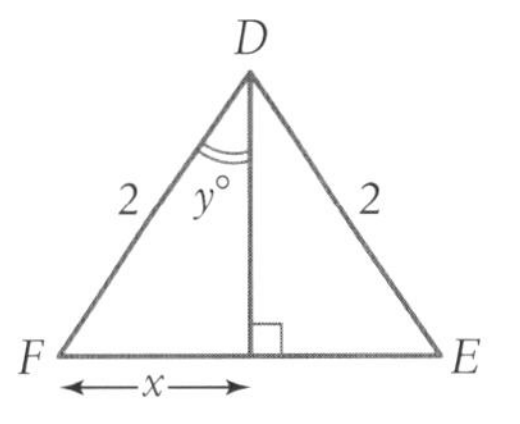

27 $\triangle PQR$ and $\triangle STU$ are similar.

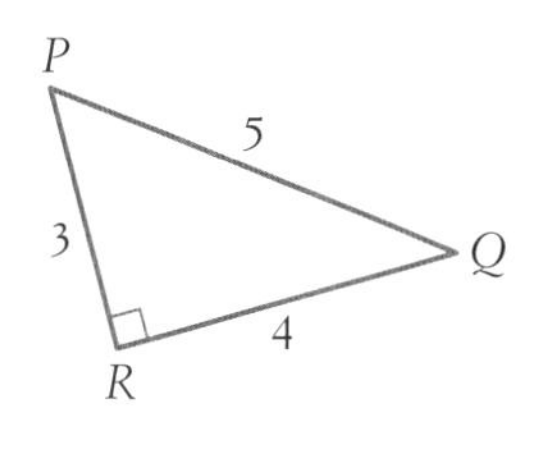

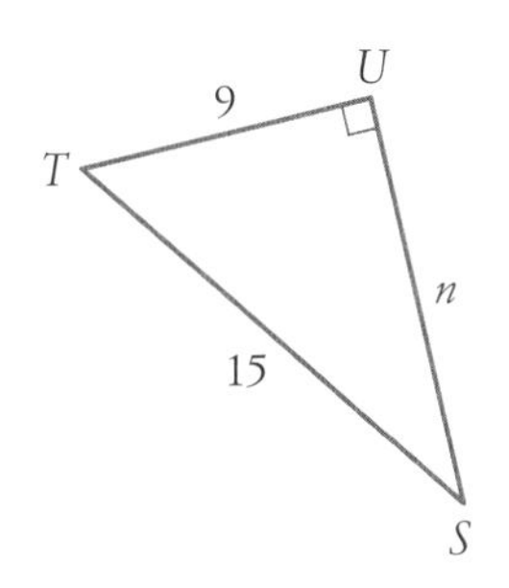

a Find the value of n.

b Is $\angle P$ equal to $\angle S$ or $\angle T$?

28 **a** Find the value of d in the diagram below.

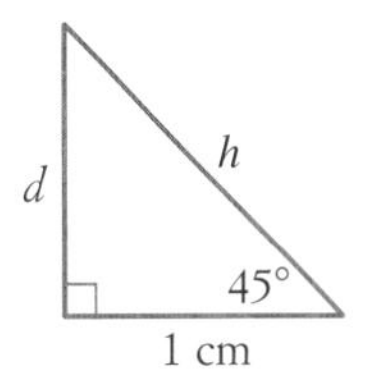

b Find the value of h, to 2 decimal places.

PART C: CHALLENGE

Bonus / 3 marks

Find an angle size, θ, which satisfies the equation $\sin\theta = \cos\theta$. Can you prove your answer?

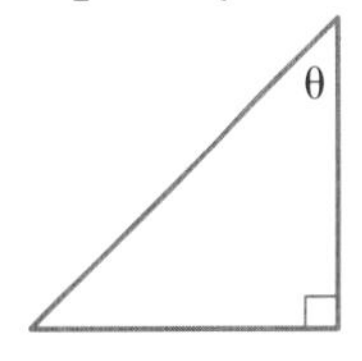

9780170454551

8 FINDING AN UNKNOWN ANGLE

FIRST DECIDE WHETHER YOU'RE GOING TO USE SIN, COS OR TAN.

For each triangle, find the size of the angle marked, correct to the nearest degree.

1

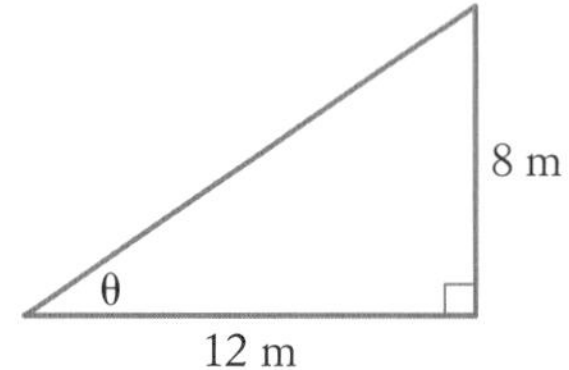

$\theta =$ ____________

2

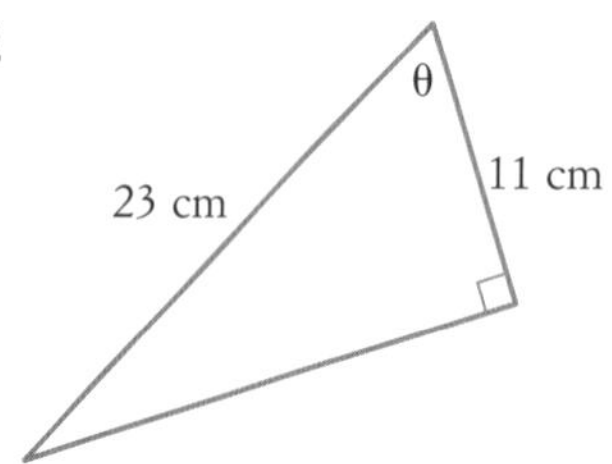

$\theta =$ ____________

3

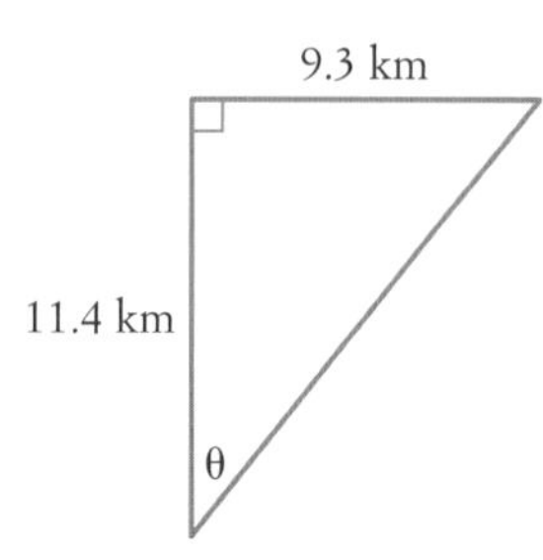

$\theta =$ ____________

4

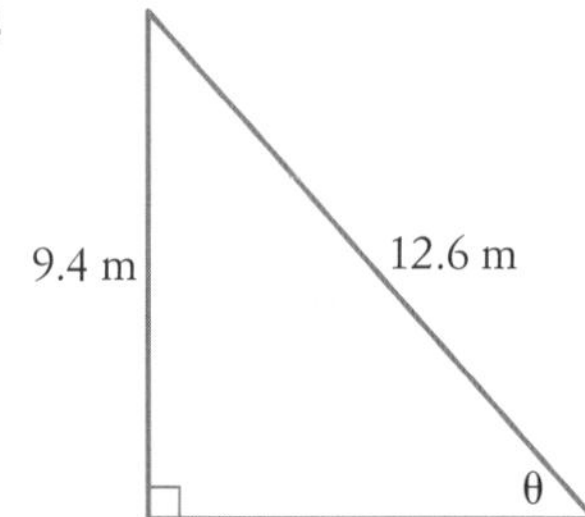

$\theta =$ ____________

5

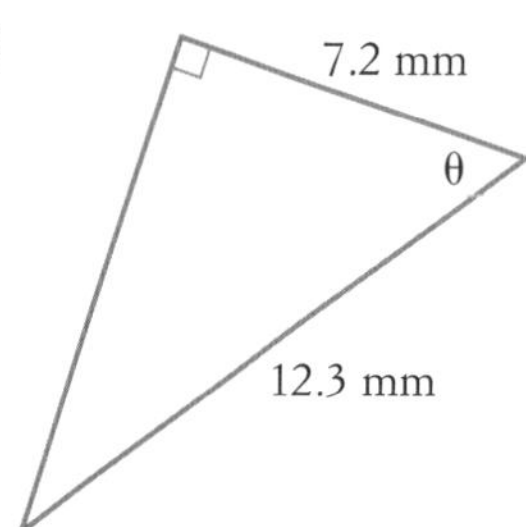

$\theta =$ ____________

6

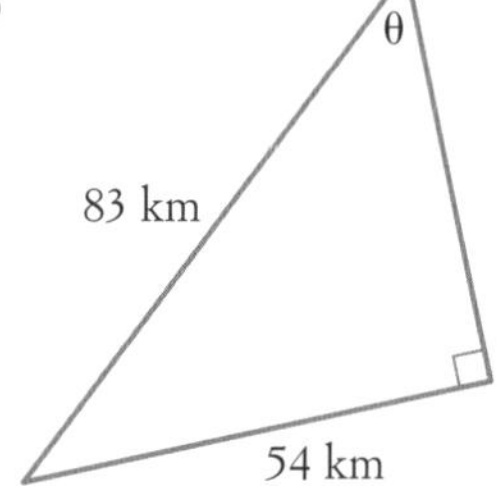

$\theta =$ ____________

7

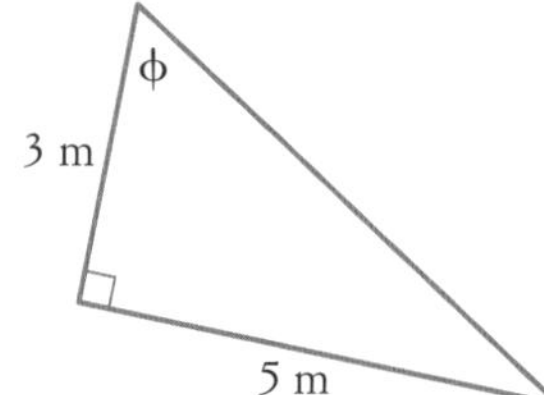

$\phi =$ ____________

8

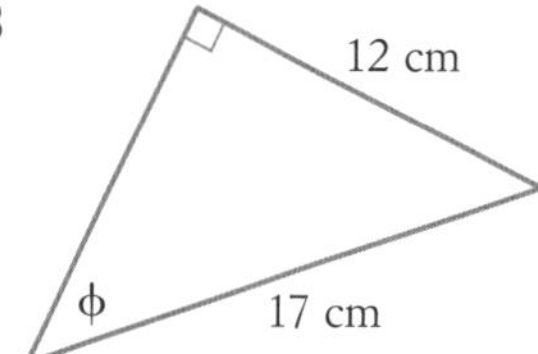

$\phi =$ ____________

9

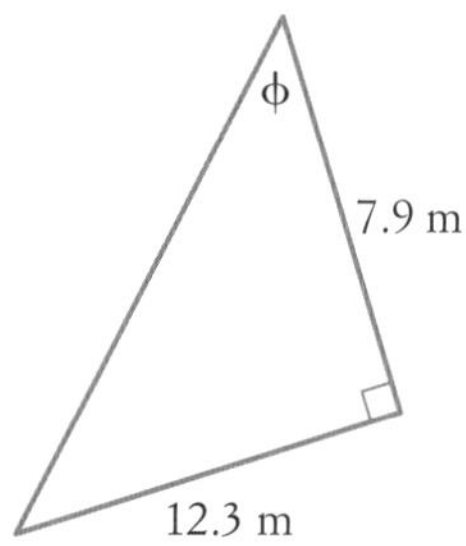

$\phi =$ ____________

10

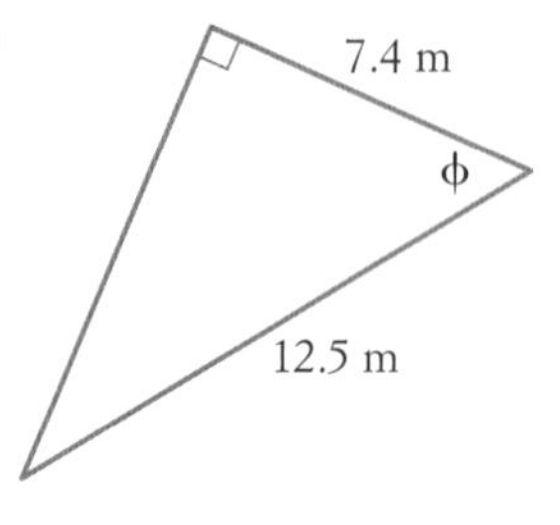

$\phi =$ ____________

11

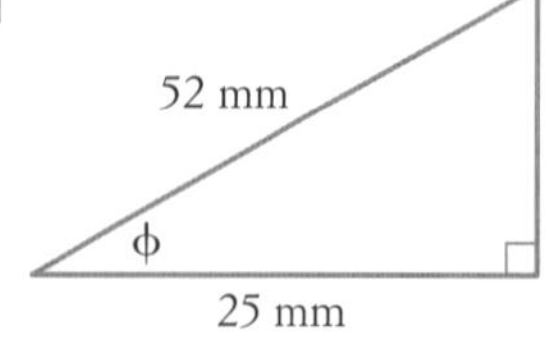

$\phi =$ ____________

12

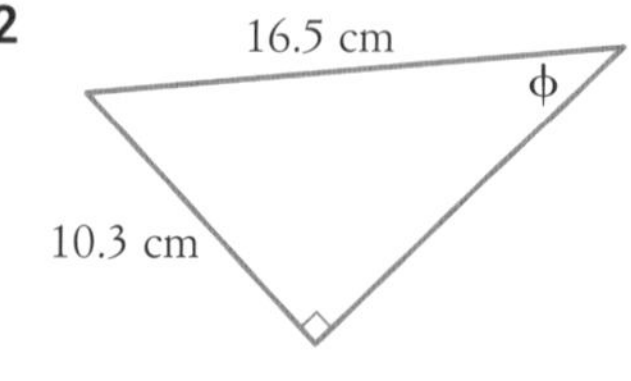

$\phi =$ ____________

13

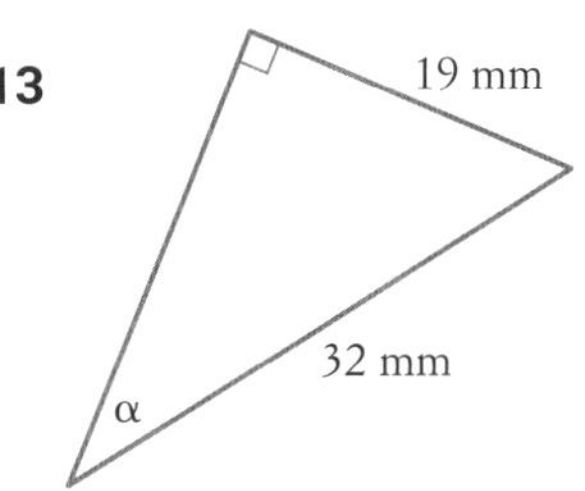

α = ________

14

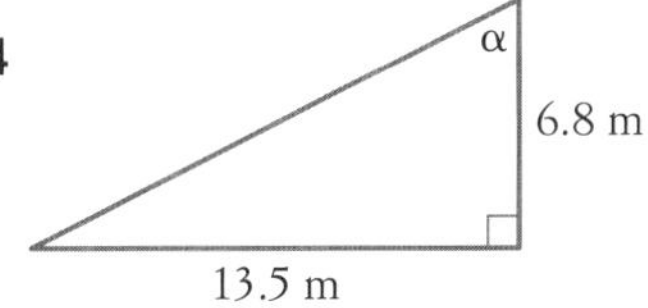

α = ________

15

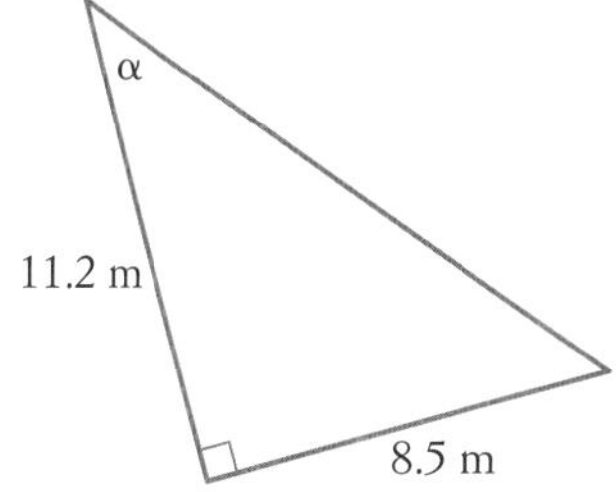

α = ________

16

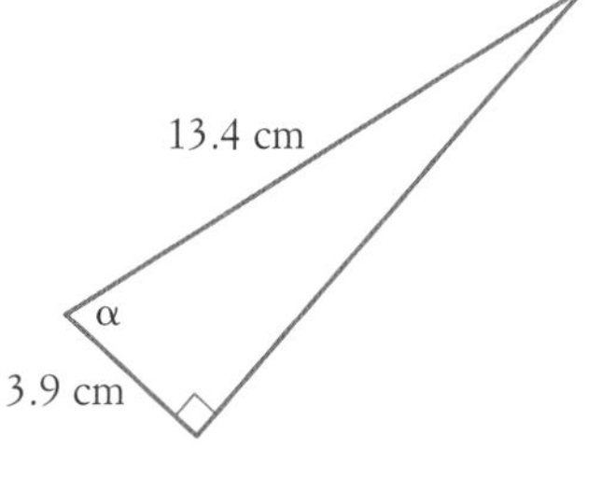

α = ________

17

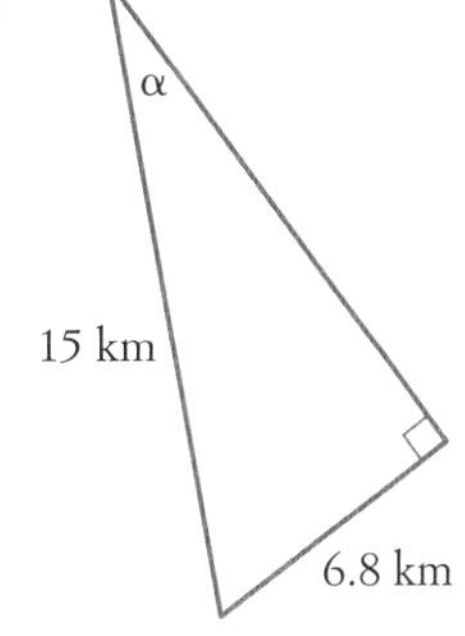

α = ________

18

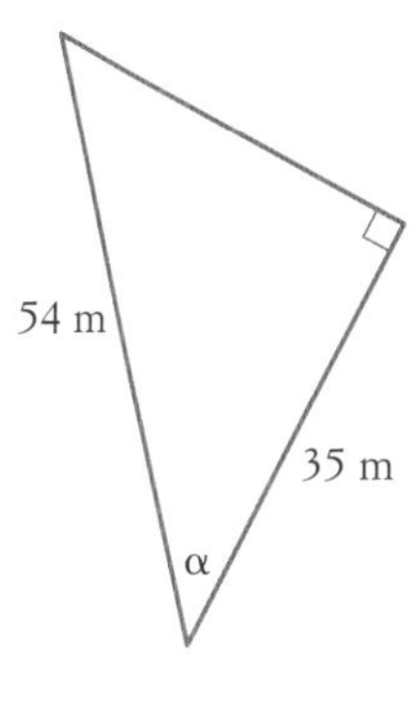

α = ________

19

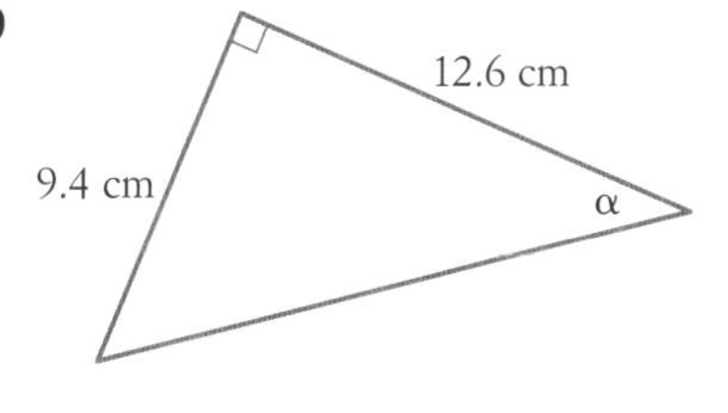

α = ________

20

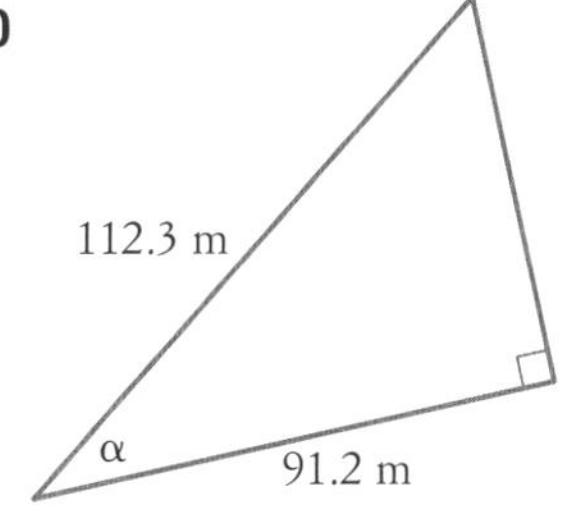

α = ________

21

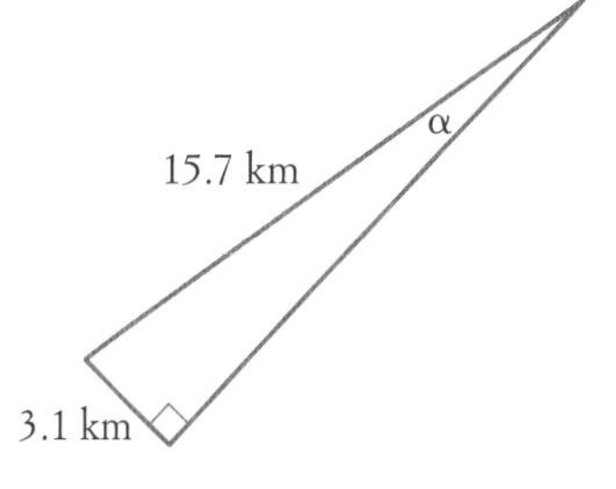

α = ________

Mixed answers: 57°, 61°, 63°, 37°, 54°, 36°, 36°, 27°, 39°, 59°, 37°, 27°, 34°, 45°, 39°, 50°, 41°, 61°, 48°, 73°, 54°, 11°

WS WORKSHEET

8 ELEVATIONS AND BEARINGS

THIS WORKSHEET HAS YOU APPLYING TRIGONOMETRY, INCLUDING ANGLES OF ELEVATION AND DEPRESSION.

WS WORKSHEET

1 A rally driver drove 10 km on a bearing of 018°. How far north did he travel, to 2 decimal places? ______

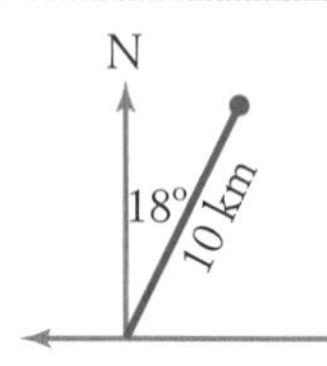

2 A skateboard ramp is 40 cm high and makes an angle of 20° with the ground. How long is the ramp, to the nearest centimetre?

______

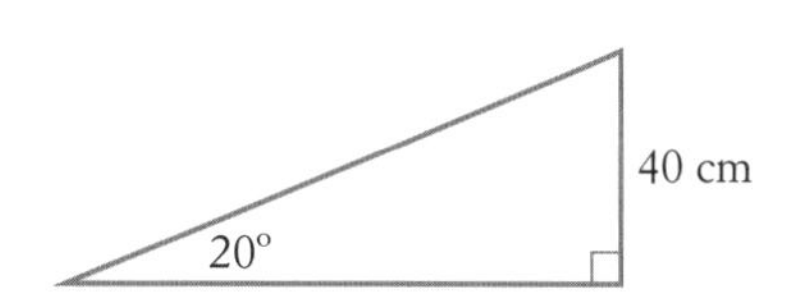

3 From the top of a 126 m tower, Kim saw her house at an angle of depression of 50.5°. How far is her house from the foot of the tower (to the nearest metre)? ______

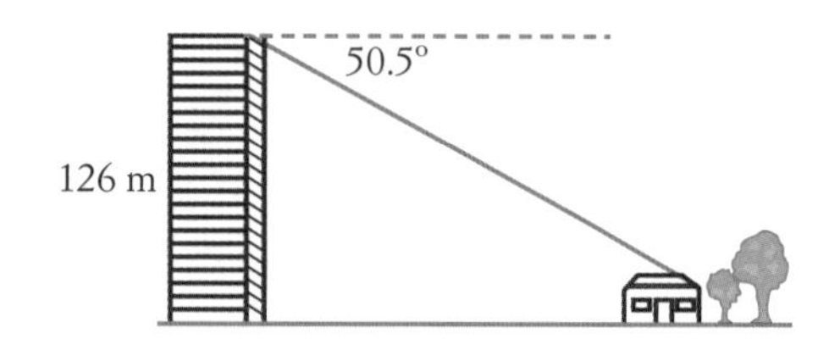

4 An aircraft took off at an angle of 32°. How far had it flown when it reached a height of 1200 m? (Answer to the nearest metre.) ______

5 A fisherman travels 16 km on a bearing of 172°. How far east is he from his starting point? Answer correct to 2 decimal places. ______

6 Gene saw an airship flying horizontally at an angle of elevation of 31° and a height of 950 m. How far, to the nearest metre, must the airship fly so that it is directly above him? ______

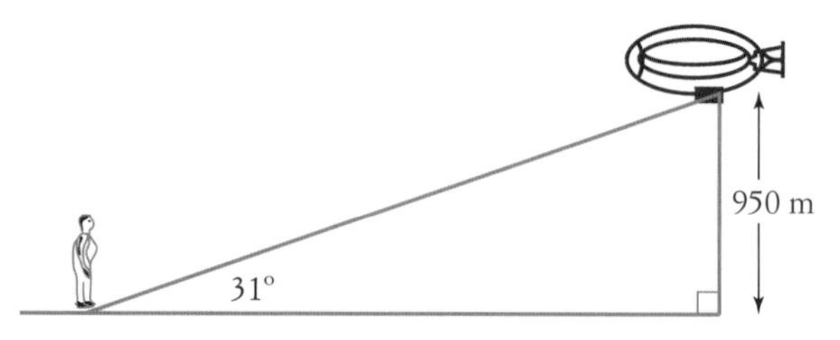

7 A lighthouse is 500 nautical miles due north of a ship. If the ship sails on a course bearing 333° until it is due west of the lighthouse, how far west is it? Answer correct to 2 decimal places.

______

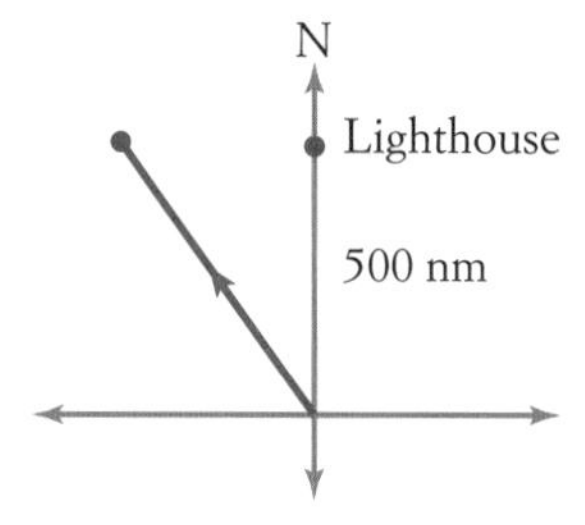

8 Bountha walked from base camp in a northwesterly direction for 6 km until he was 4.3 km north of camp. What is his bearing from camp? ______

9 The Katoomba scenic railway is 450 m long and has a vertical drop of 250 m. What angle, to the nearest minute, does it make with the horizontal? ______

C S F

9780170454551

10 From the top of a building 90 m high, Mili saw a ferry on Sydney Harbour that was 400 m from the base of the building. What was the angle of depression of her line of sight? Answer correct to 2 decimal places. ______________

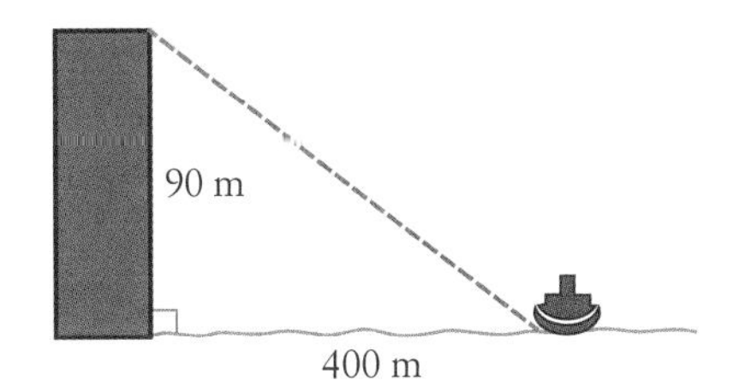

11 A speedboat travelled on a steady course bearing 210° from P to Q. If Q is 80 nautical miles further west than P, find how far the boat travelled.

____

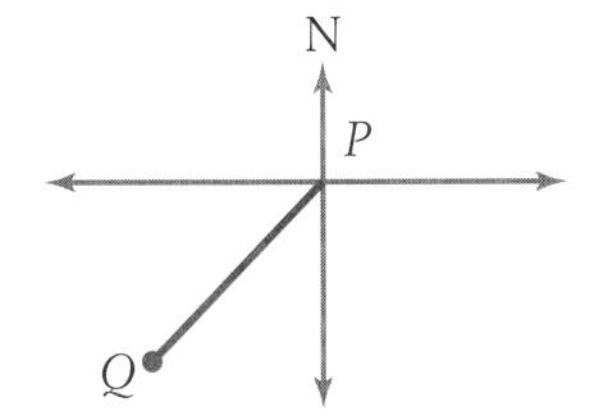

12 Macarthur Park has a slippery slide inclined at an angle of 40°, and the bottom of the slide is 5 m from the base of the ladder.

Strawberry Field has a slide inclined at 50°, and the bottom is 4 m from the base of the ladder.

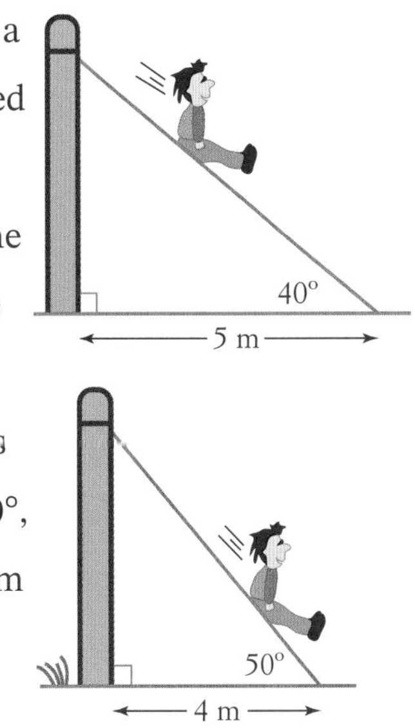

a Which park has the higher slide?

____

b Which park has the longer slide?

____

8 TRIGONOMETRY CROSSWORD

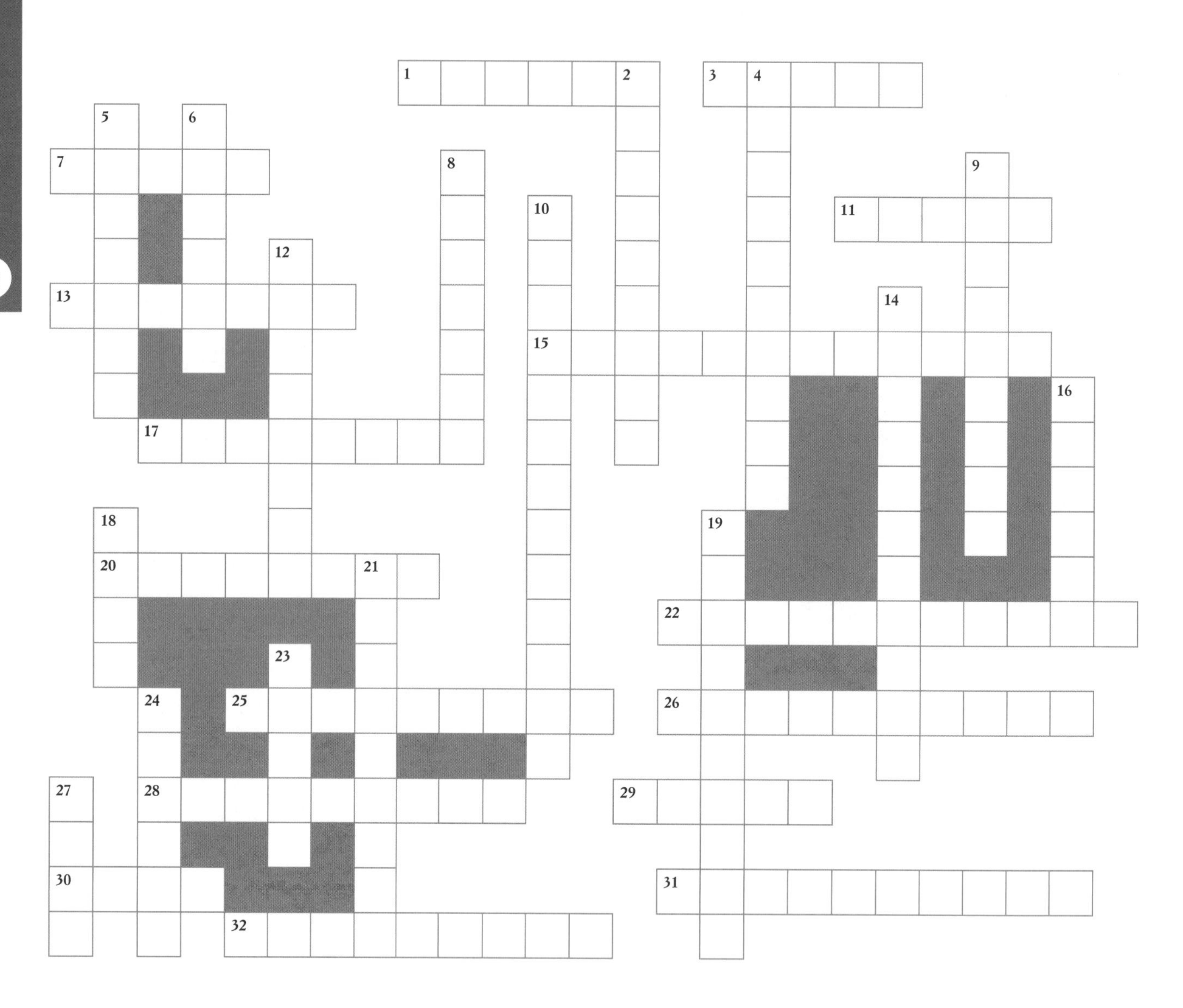

PS PUZZLE SHEET

9780170454551

Clues across

1 $\frac{1}{60}$ of a degree

3 The number of digits in a bearing

7 $\frac{\text{opposite}}{\text{adjacent}}$ is an example of a trigonometric r ______

11 A measure of turning

13 An angle that shows direction

15 The mathematics of triangle measurement

17 The side facing the angle

20 The side next to the angle and pointing to the right angle in a right-angled triangle

22 The hypotenuse is the longest side of this type of triangle

25 This compass direction has bearing 045°

26 Opposite of vertical

28 Halfway between west and south

29 The Greek letter θ

30 $\frac{\text{opposite}}{\text{hypotenuse}}$

31 Angle of 'looking down'

32 The direction of Sydney from Darwin

Clues down

2 Opposite of depression

4 $\cos = \frac{\text{adjacent}}{?}$

5 The full name of tan

6 230° is an example of a three-______ bearing

8 $\sin^{-1}$ means 'i ______ sine'

9 Angles between parallel lines that are equal

10 This compass direction has bearing 247.5°

12 'M' word meaning memory aid for learning

14 The bottom half of a fraction

16 A unit for measuring angles

18 Where does the Sun rise?

19 Instrument for measuring angle of elevation

21 The opposite of southeast

PS PUZZLE SHEET

8 TRIGONOMETRY 1

Name:

Due date:

Parent's signature:

Part A	/ 8 marks
Part B	/ 8 marks
Part C	/ 8 marks
Part D	/ 8 marks
Total	/ 32 marks

PART A: MENTAL MATHS

Calculators not allowed

1 Convert each percentage to a decimal.

a 82% ____________________

b $7\frac{1}{2}\%$ ____________________

2 Find d.

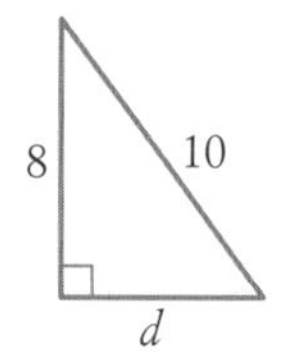

____

3 Find the value of x and y.

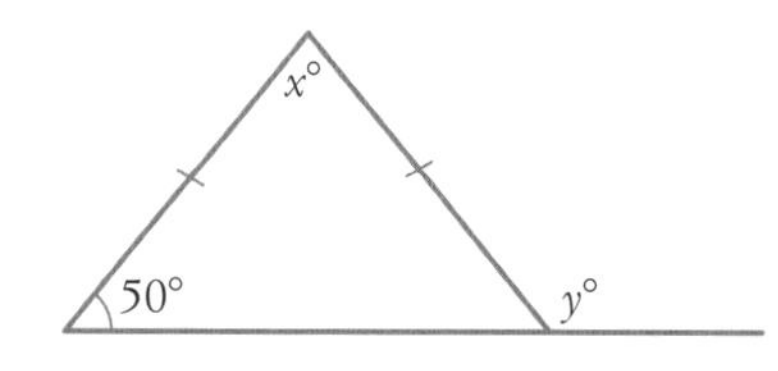

____

____

4 Simplify:

a $(7c)^0$ ____________________

b $\left(\frac{2x}{3p}\right)^{-2}$ ____________________

5 A temperature in degrees Celsius (°C) can be converted to degrees Fahrenheit (°F) using the formula $F = \frac{9C}{5} + 32$. Convert −20°C to °F.

____

____

PART B: REVIEW

1 Find x and y, correct to 2 decimal places.

a

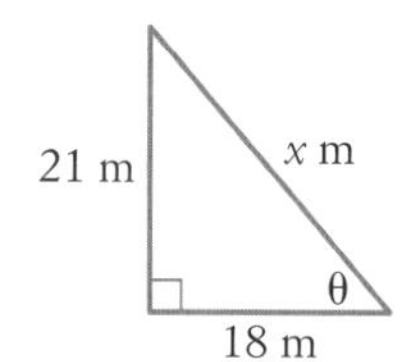

____

b

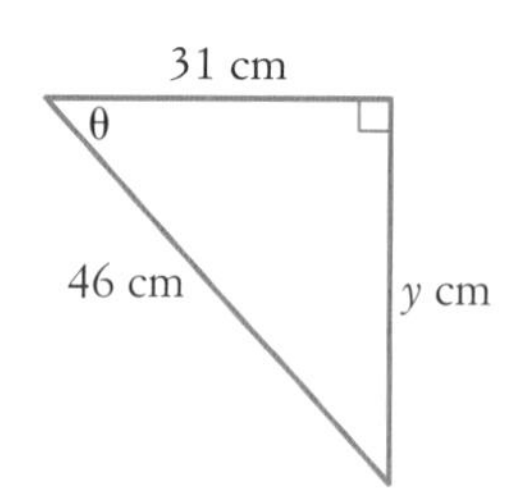

____

2 Find the size of θ in each triangle above, correct to the nearest degree.

a ____________________

b ____________________

C S F

9780170454551

3 (4 marks) Test whether each triangle is right-angled, giving reasons.

a 17, 8, 15

b 6, 23, 25

4 If $\sin A = \frac{77}{85}$, find $\tan A$ as a fraction.

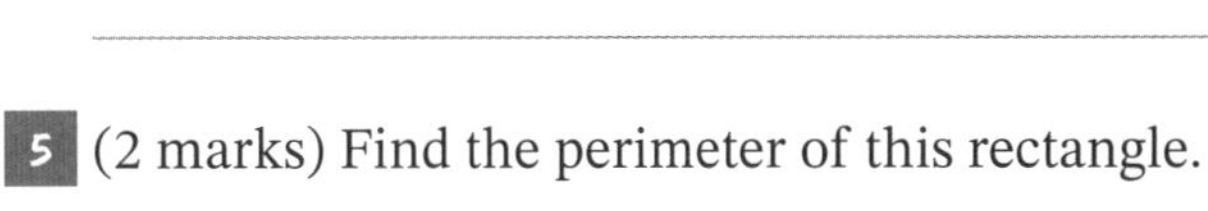

5 (2 marks) Find the perimeter of this rectangle.

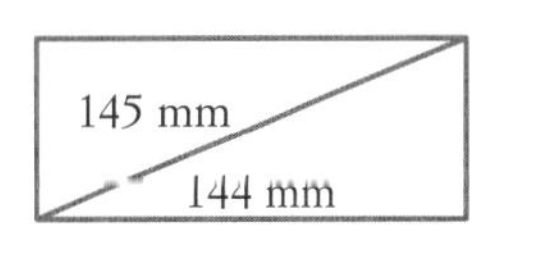

PART C: PRACTICE

› Finding an unknown side
› Finding an unknown angle

1 Evaluate 26 tan 41° 7' correct to 2 decimal places.

2 Convert 80.15° to degrees and minutes.

3 Find the value of each variable, correct to 2 decimal places.

a

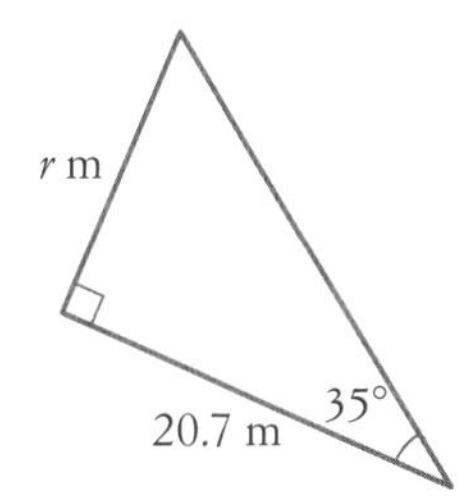

b

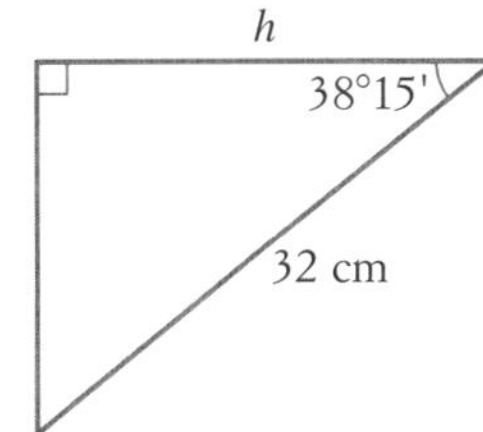

c

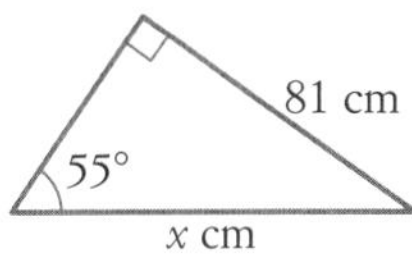

PART D: NUMERACY AND LITERACY

1 Write the trigonometry meaning of:

a hypotenuse

b tangent

c minute

2 Find the size of angles A and C, correct to the nearest minute.

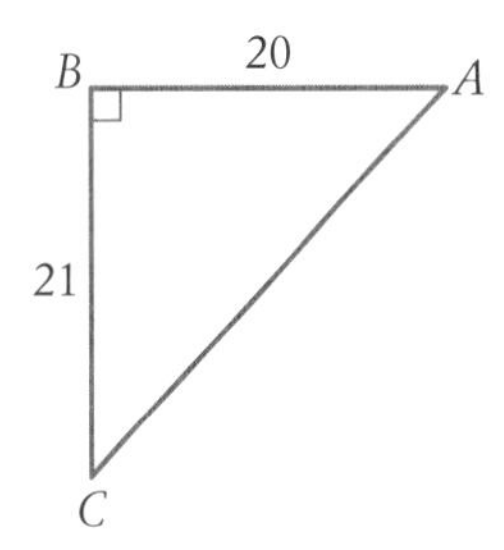

3 (3 marks) In $\triangle ABC$, $B = 90°$, $AB = 18$ cm and $A = 67.2°$. Draw a diagram and find, correct to one decimal place, the length of:

a AC

b BC

HW HOMEWORK

8 TRIGONOMETRY 2

BEARINGS ARE MEASURED FROM NORTH IN A CLOCKWISE DIRECTION, FROM 000° TO 360°.

Name:

Due date:

Parent's signature:

Part A	/ 8 marks
Part B	/ 8 marks
Part C	/ 8 marks
Part D	/ 8 marks
Total	/ 32 marks

PART A: MENTAL MATHS

Calculators not allowed

1 (4 marks) Solve each equation.

a $\frac{7-a}{3} = \frac{3a+2}{4}$

b $\frac{5x}{6} + \frac{3x}{5} = 5$

2 Increase \$120 by 20%.

3 Find x and y for this pair of similar triangles.

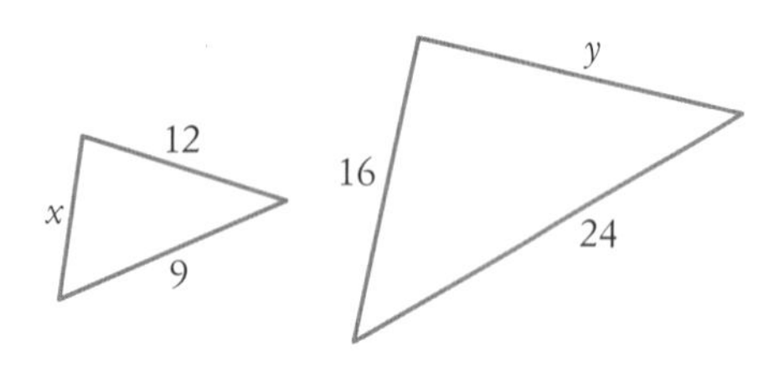

4 Write 0.000 020 7 in scientific notation.

PART B: REVIEW

1 Find the value of each variable, correct to 2 decimal places.

a

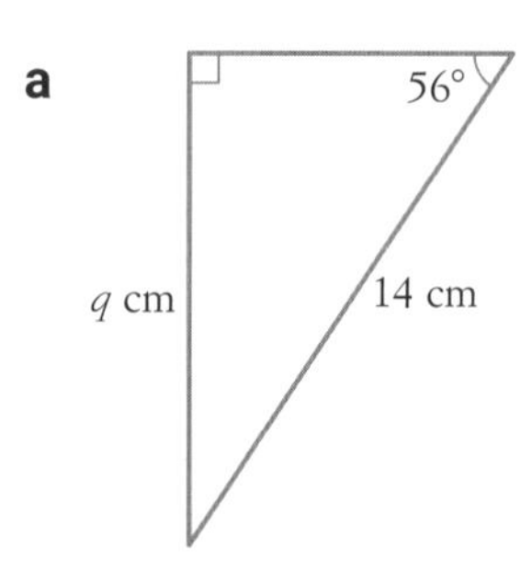

b

8 mm

b mm

29°

2 (2 marks) A supporting wire is attached to the top of a flagpole. The wire meets the ground at an angle of 48° and the flagpole is 13 m high. How far from the base of the flagpole is the wire anchored to the ground, correct to one decimal place?

C S F

HW HOMEWORK

9780170454551

3 Find θ, correct to the nearest degree.

a

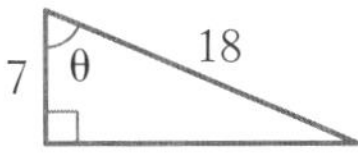

b

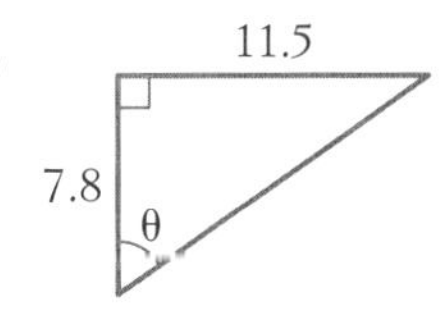

4 (2 marks) $\triangle GHI$ is right-angled at H, $HI = 15.7$ cm and $HG = 9.3$ cm.

Find $\angle I$, correct to the nearest minute.

PART C: PRACTICE

› Angles of elevation and depression
› Bearings

1 From 23 m away, Ashley sees the top of a building at an angle of elevation of 27°. Find the height of the building, correct to one decimal place.

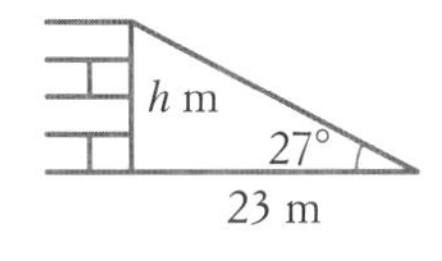

2 Find the bearing of each point from O.

a

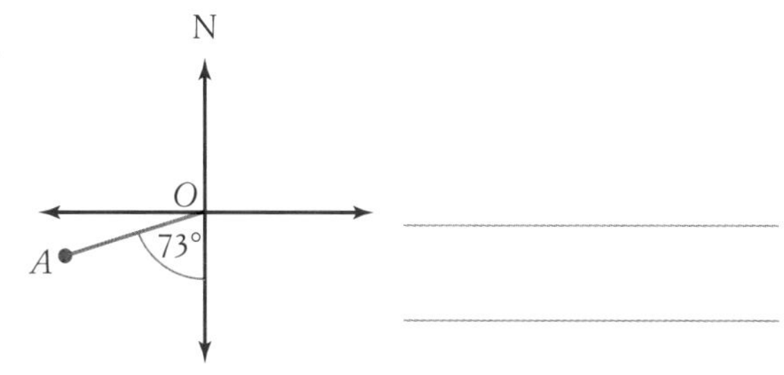

b

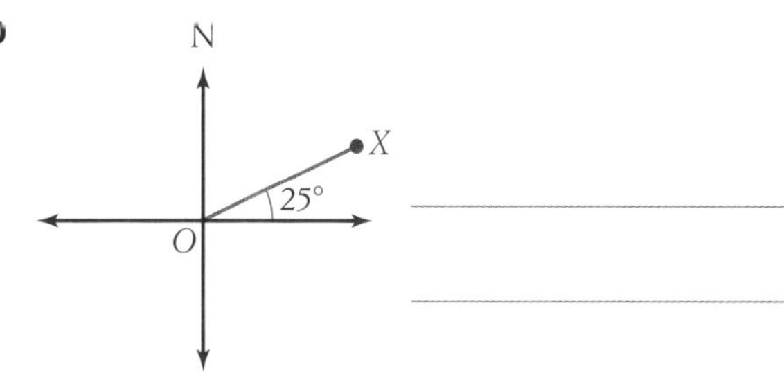

c

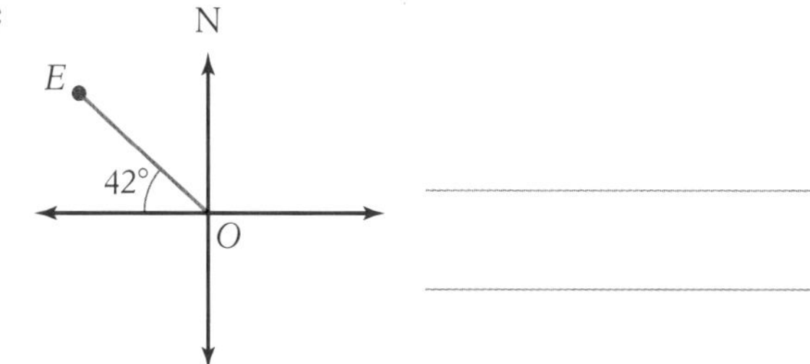

3 (4 marks) Kasun sailed from port on a bearing of 200° for 6 km.

a How far south (to one decimal place) was he from port?

b What is the bearing of port from his boat?

C S F

HW HOMEWORK

PART D: NUMERACY AND LITERACY

1 Complete:

a An angle of depression is the angle of looking __________ at an object, measured from the __________ to the line of sight.

b __________ -figure bearings, also called __________ bearings, show direction measured from North in a __________ direction.

2 Write the bearing of southwest.

____

3 (2 marks) Find the angle of depression (to the nearest degree) of a boat that is 180 m from the base of a 62 m cliff.

____

____

HW HOMEWORK

C S F

9780170454551

A PAGE OF BEARINGS 8

WORKSHEET

1

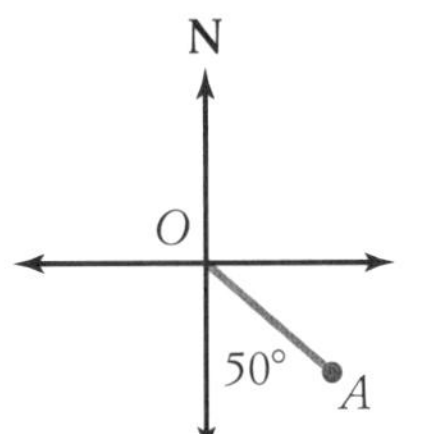

2

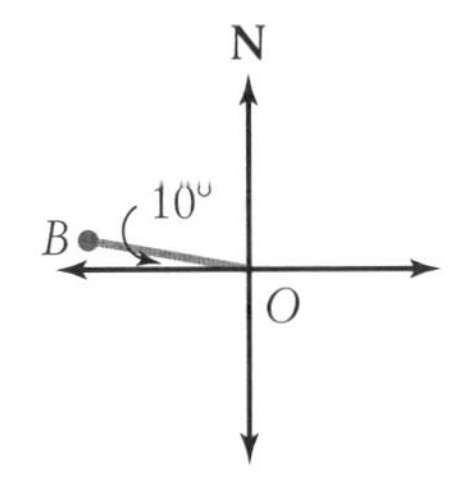

3

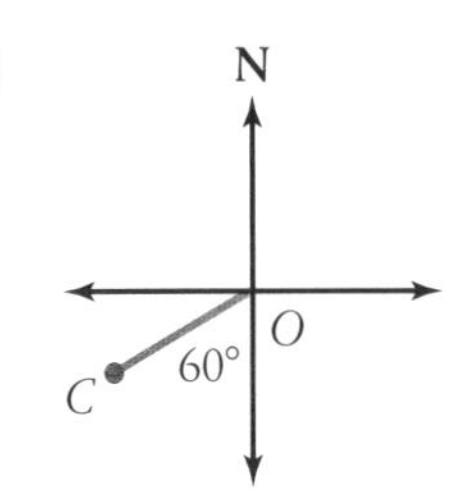

4

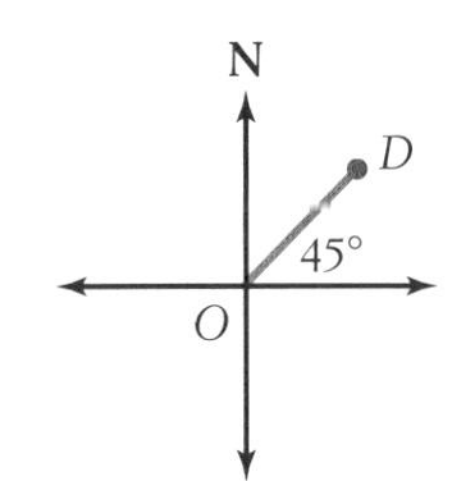

5

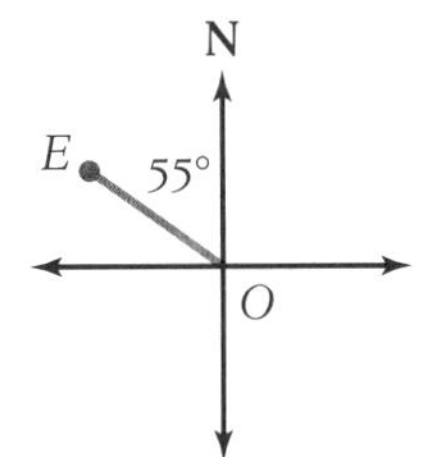

6

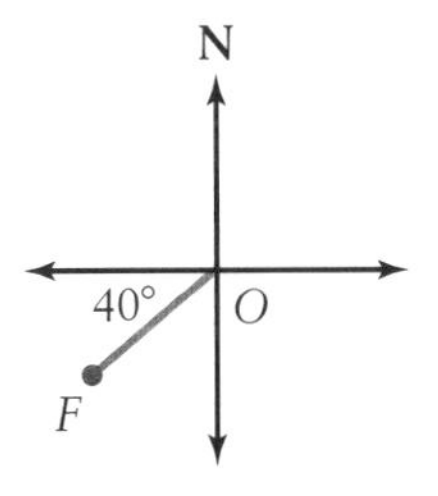

7

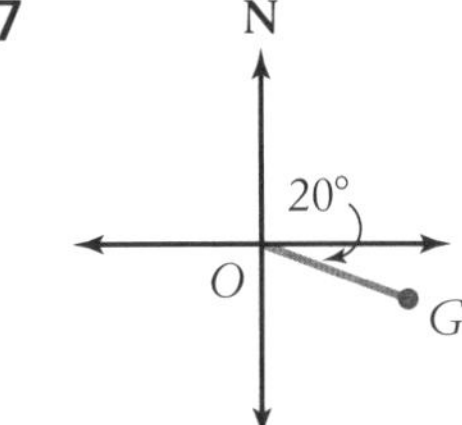

8

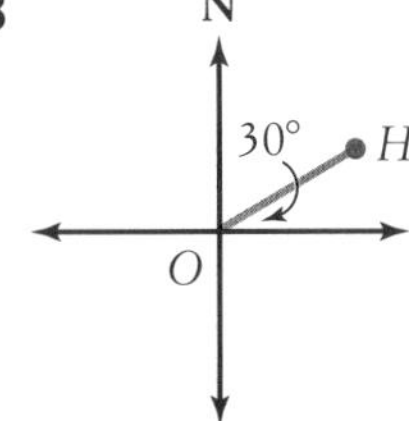

9

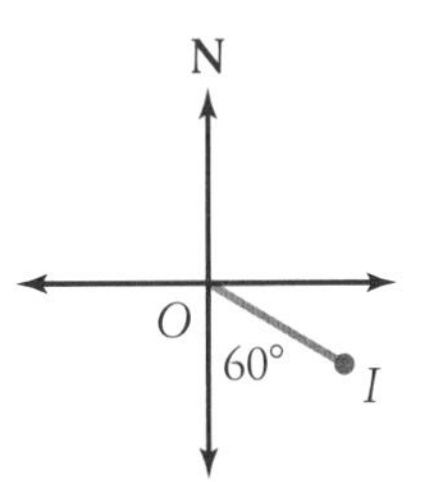

10

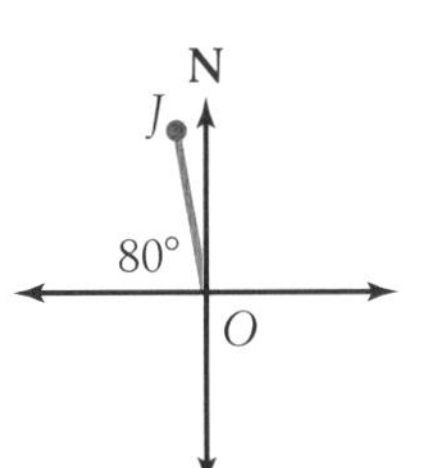

11

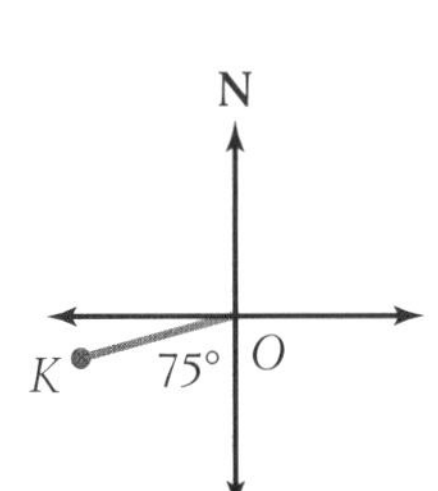

12

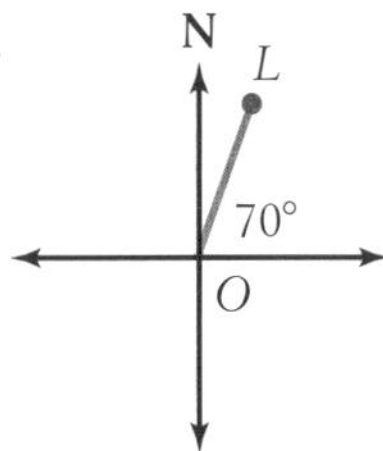

13

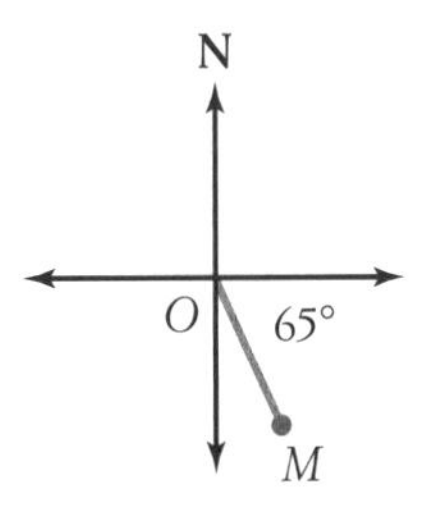

14

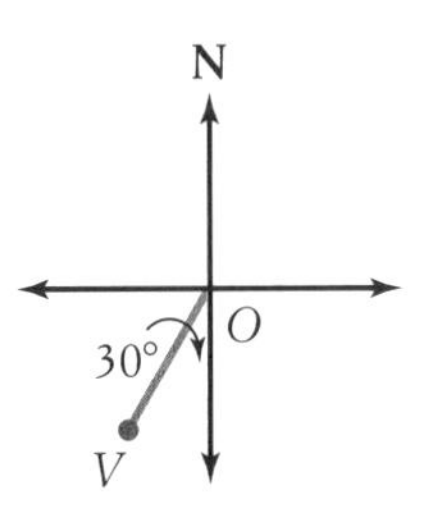

15

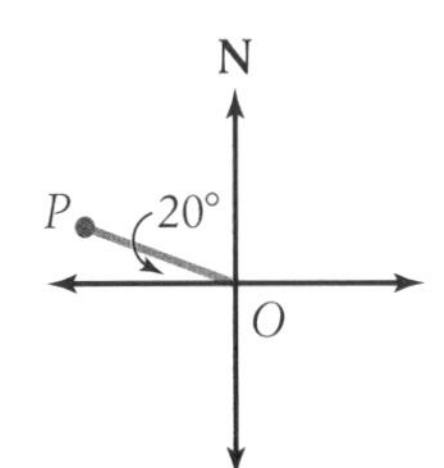

16

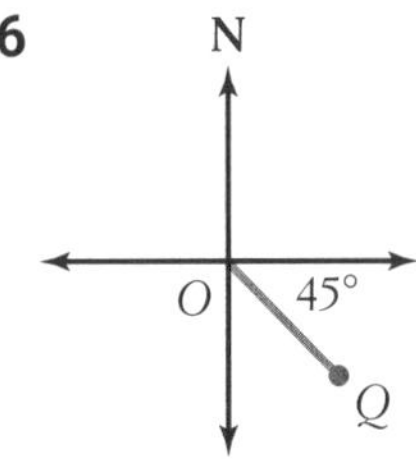

17

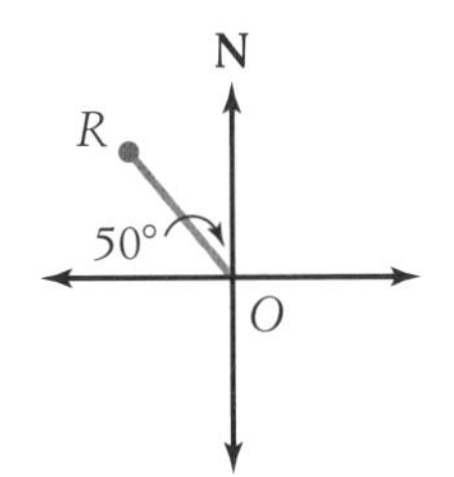

18

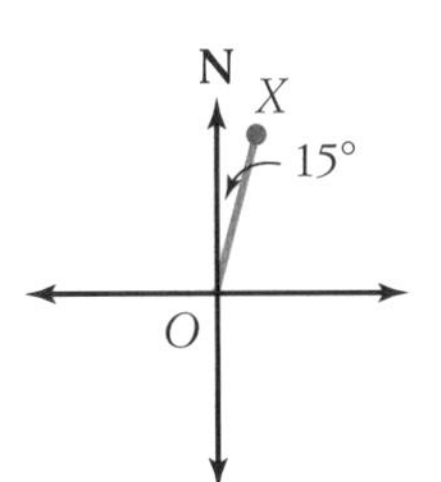

19

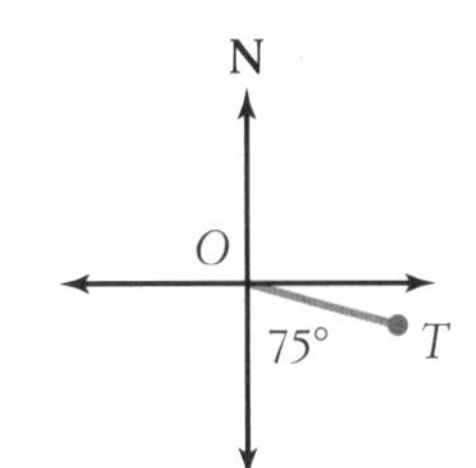

20

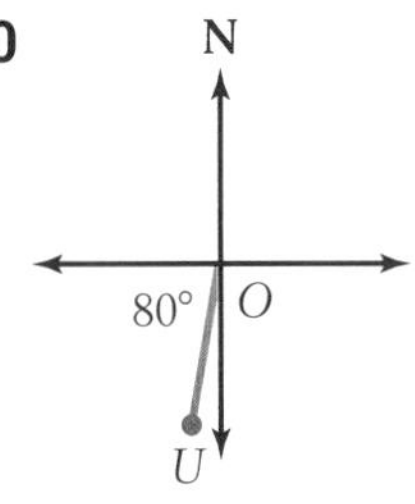

Mixed answers: 120°, 130°, 350°, 015°, 280°, 155°, 110°, 310°, 020°, 305°, 255°, 045°, 190°, 290°, 135°, 240°, 230°, 060°, 210°, 105°

WS WORKSHEET

9 STARTUP ASSIGNMENT 9

PART A: BASIC SKILLS / 15 marks

1 Evaluate $\sqrt{150}$ correct to 3 significant figures. ______

2 Calculate Arden's fortnightly pay if he earns a salary of \$85 300. ______

3 Write an expression for the next even number after n if n is odd. ______

4 What is this formula used for?

$A = \frac{1}{2}xy$ ______

5 Prove that this triangle is right-angled.

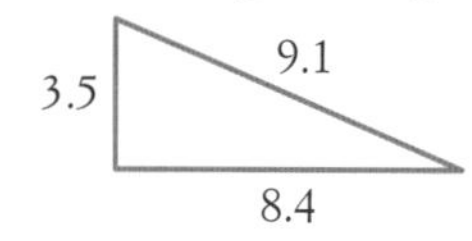

6 Evaluate:

a $4^3 \times 5^0$ ______

b 3^{-2} ______

7 Simplify:

a $4(x + 3) - 2(x - 3)$ ______

b $\frac{2p}{3} \times \frac{3p}{10}$ ______

8 Find the area of this sector, correct to 2 decimal places.

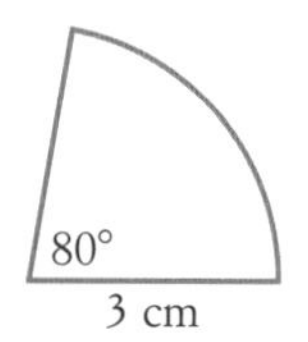

______

9 Express 2.8×10^{-5} in decimal form.

______

10 Solve $\frac{2x + 7}{3} = 4$. ______

______

11 List the 4 tests for congruent triangles.

______

12 The mean of 13, 19, 12, x and 17 is 16. Find x.

______

______

13 Find the curved surface area of this cylinder, correct to 2 decimal places.

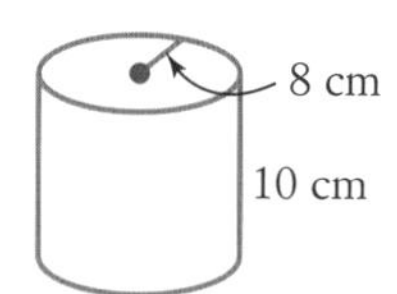

______

PART B: ALGEBRA AND EQUATIONS / 25 marks

14 Simplify each expression.

a $-8(5g - 4)$ ______

b $3(2y - 10) + 7y$ ______

c $-4(8 + 5a) + 2(a + 13)$ ______

______

d $9(7b - 3) - 4(10 - 3b)$ ______

______

15 Solve each equation.

a $3p = -15$ ______

b $2f + 13 = 30$ ______

c $14 - 4d = 20$ ______

d $2.5g - 16 = 32$ ______

C S F

9780170454551

e $\dfrac{x + 5}{3} = 6$

f $4a - 3 = a + 8$

16 Find y when $x = 2$.

a $y = 7x + 15$

b $y = 4(17 - 3x)$

c $5x + 2y = 24$

17 Find a when $b = -5$.

a $14 + 8b = a$

b $a = 45 - 9b$

c $5(b + 12) = a$

18 Test whether $x = 3$ is a solution to:

a $50 - 11x = 17$

b $2x + 2 = 20 - 4x$

c $6(x - 5) = 12$

19 Graph each line on the same number plane.

a $y = 4$ b $x = 5$

c $y = 2x - 3$ d $y = -x + 5$

20 Use an equation to find 2 consecutive even numbers that add to 150.

21 Use an equation to find 3 consecutive odd numbers that add to 105.

PART C: CHALLENGE

Bonus / 3 marks

Robbie Rabbit and Timmy Turtle had a race. Robbie ran at a speed of 10 km/h for half the distance, then 8 km/h for the other half. Timmy ran 9 km/h for the entire distance. Who won the race, Robbie or Timmy?

C S F

WS WORKSHEET

9 INTERSECTION OF LINES

Graph and label each pair of linear functions on the number plane below. It may help to plot each pair of lines with a different colour. Write the point of intersection of each pair of lines.

1 $y = x + 2$

$y = 3x - 4$

(____, ____)

2 $y = 2x - 2$

$y = -x + 1$

(____, ____)

3 $y = -x + 2$

$y = -\frac{1}{4}x - 1$

(____, ____)

4 $y = \frac{1}{2}x + 7$

$y = -3x$

(____, ____)

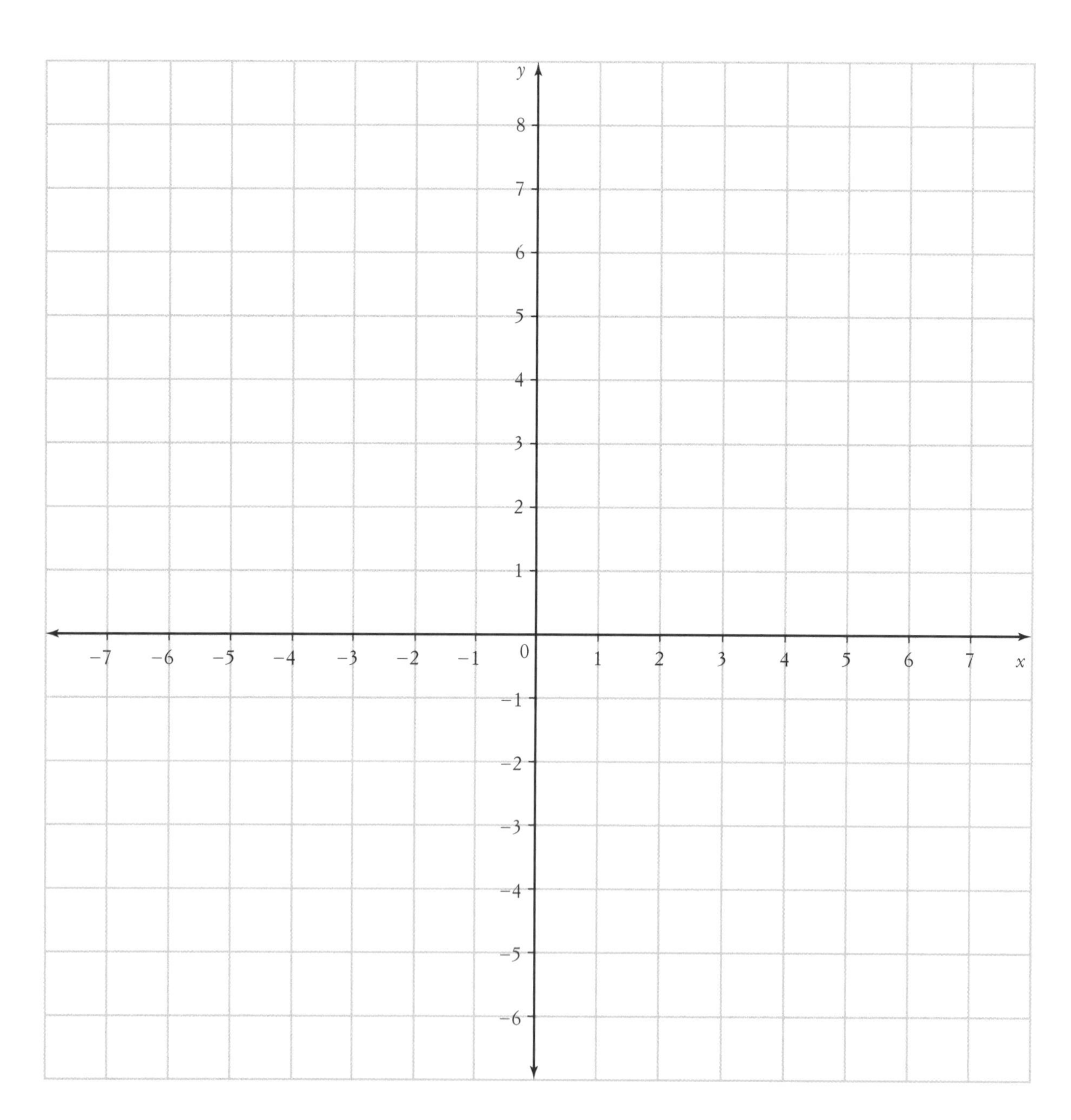

9780170454551

SIMULTANEOUS EQUATIONS CROSSWORD 9

UNSCRAMBLE THE KEYWORDS BELOW AND PLACE THEM IN THE CROSSWORD.

PUZZLE SHEET

Clues across

7 LOVES

11 NORENTCITIES

12 OPTIN

14 STAYSIF

15 IBISSTUNTOUT

16 EDMOTH

17 OUSTLION

Clues down

1 EMAILSUNTOUS

2 LOBPERM

3 BRAVELAI

4 ECLAIRBAG

5 EXSA

6 HAPGARLIC

8 SOANTIQUE

9 IMINITALONE

10 ENLIAR

13 NIECETICOFF

9 SIMULTANEOUS EQUATIONS 1

Name:

Due date:

Parent's signature:

Part	Marks
Part A	/ 8 marks
Part B	/ 8 marks
Part C	/ 8 marks
Part D	/ 8 marks
Total	/ 32 marks

HW HOMEWORK

PART A: MENTAL MATHS

Calculators not allowed

1 Evaluate $\sqrt{225}$. ______

2 Classify each type of data as categorical or numerical. If numerical, then classify as discrete or continuous.

a The number of people at a meeting.

______

b The amount of rain in Canberra today.

______

c The suburb of your home.

______

3 Find the area of each shape as a simplified algebraic expression.

a

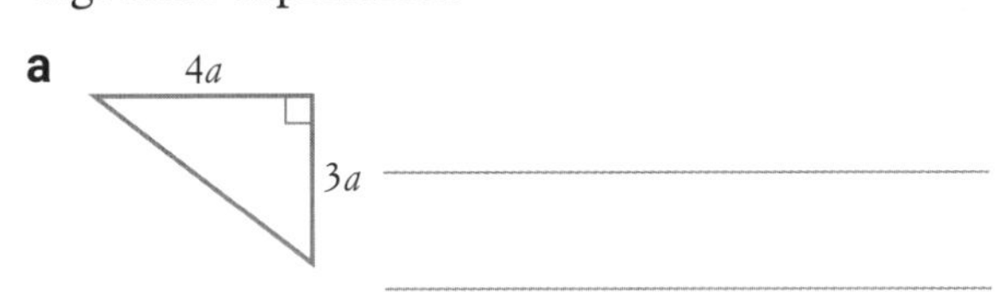

______

b

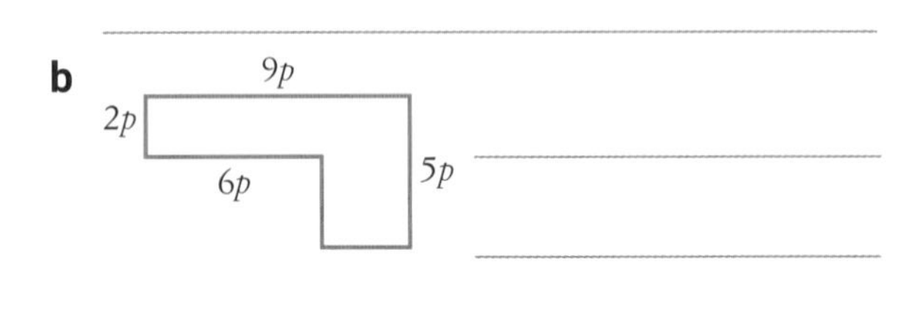

______

4 Find the mode of these data values.

Score	Frequency
13	4
14	6
15	7
16	8

______

5 Simplify $(2x^2y^3)^3$. ______

PART B: REVIEW

1 For $y = 5x + 3$, find y when:

a $x = -2$

______

b $x = -0.1$

______

2 (2 marks) Complete the table of values for $x - y = 2$, then graph the equation on the number plane.

x	−1	0	1	2
y				

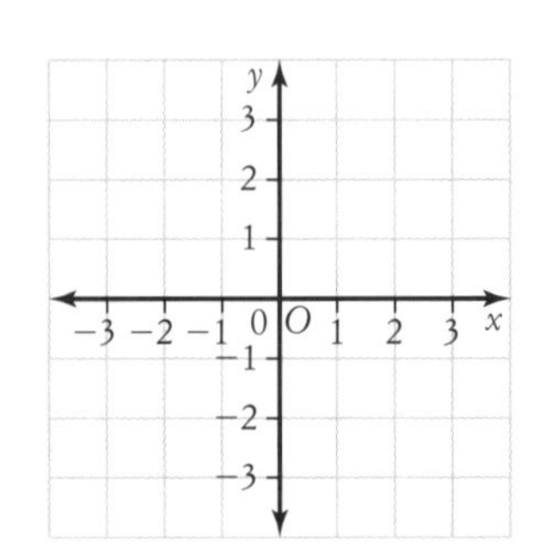

C S F

9780170454551

3 Test whether the point $(-1, -1)$ lies on the line:

a $y = 3 - 2x$

____

b $6x - y = -5$

____

4 Write the gradient and y-intercept of each linear equation, then graph the equation.

a $y = \frac{3}{5}x - 3$

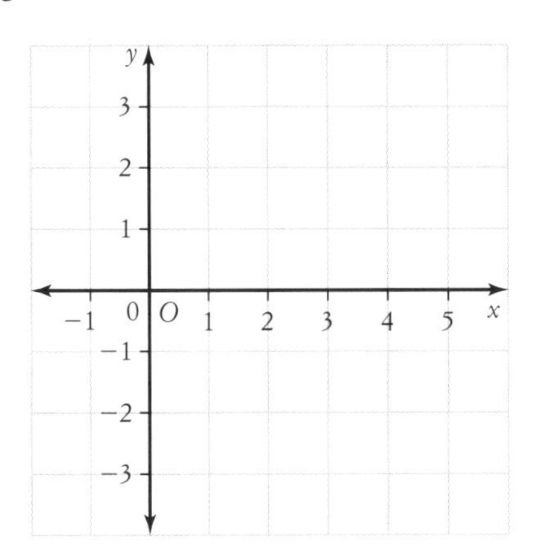

b $y = -\frac{1}{2}x + 3$

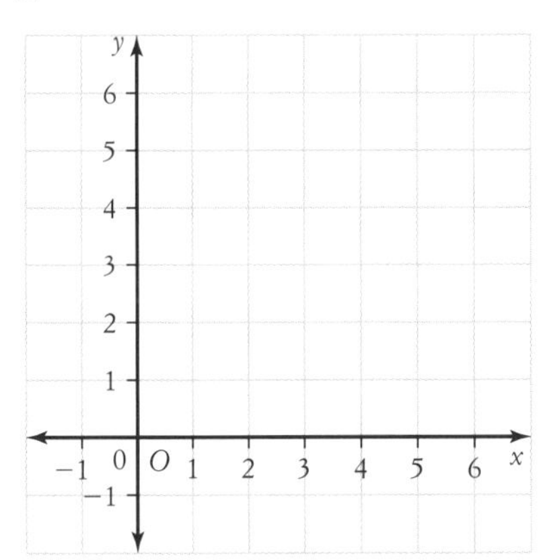

PART C: PRACTICE

› Graphical solution
› The elimination method

1 (4 marks) Complete the table of values for both equations, then solve the 2 simultaneous equations graphically.

$2x + y = 3$

x	0	1	2	3	4
y					

$x + y = -1$

x	0	1	2	3	4
y					

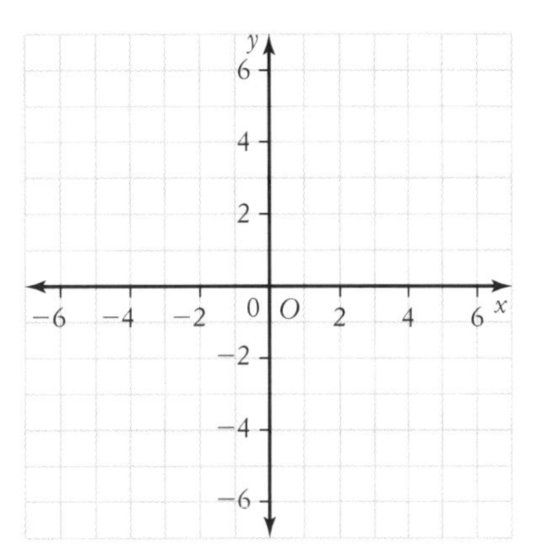

$x =$ ________, $y =$ ________

2 (4 marks) Solve these simultaneous equations using the elimination method.

$$8m - 5n = -21$$
$$4m + 5n = 45$$

PART D: NUMERACY AND LITERACY

1 (2 marks) What are the 2 algebraic methods of solving simultaneous equations called?

2 How do you find the solution to simultaneous equations using their graphs?

3 How can you check the solution of simultaneous equations?

4 (4 marks) Solve these simultaneous equations using the elimination method.

$$2x - 3y = 9$$
$$5x + y = 14$$

HW HOMEWORK

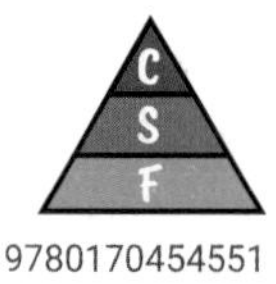

9780170454551

Name:

Due date:

Parent's signature:

Part	Marks
Part A	/ 8 marks
Part B	/ 8 marks
Part C	/ 8 marks
Part D	/ 8 marks
Total	/ 32 marks

SIMULTANEOUS EQUATIONS 2 (9)

PART A: MENTAL MATHS

Calculators not allowed

1 Round 103°45'40" to the nearest minute.

2 Simplify $(-2x^2y^0)^3$.

3 Expand $(a + 5)(2a + 3)$.

4 Find the surface area of this rectangular prism.

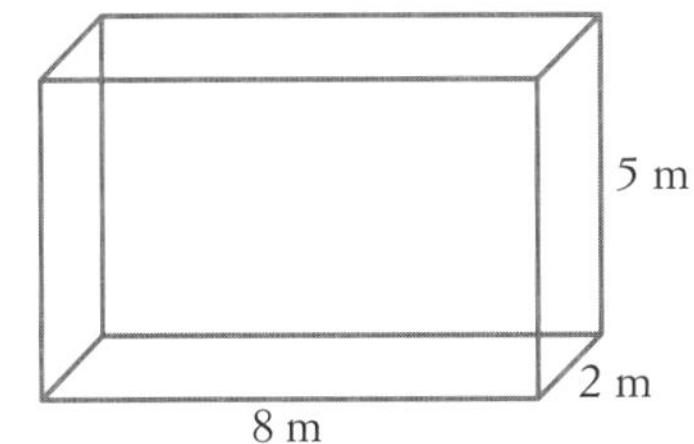

5 Calculate the time difference between 9:15 a.m and 5:20 p.m.

6 Simplify each ratio.

a 5 mm : 2 m

b 3 h : 10 min

7 Factorise $4y^2 - 24y + 32$.

PART B: REVIEW

Solve each pair of simultaneous equations using the elimination method.

1 $2a + b = 23$

$3a - 2b = 3$

2 $3m - 2n = 19$

$4m + 7n = 6$

HW HOMEWORK

9780170454551

PART C: PRACTICE

› The substitution method
› Simutaneous equations problems

1 (4 marks) Solve this pair of simultaneous equations using the substitution method.

$y = 10 + 3x$

$y = -x + 6$

2 (4 marks) It costs 4 adults and 3 children $500 for tickets to a show, while the cost of 3 adults and 2 children is $360. Use simultaneous equations to find the cost of each adult and child ticket.

PART D: NUMERACY AND LITERACY

1 When solving simultaneous equations, Melody made y the subject of one of the equations. Which method of solving simultaneous equations was she using?

2 (3 marks) Put in order the following steps for solving a problem requiring simultaneous equations.

A Solve the problem by answering in words.

B Solve the equations.

C Write simultaneous equations for the problem.

D Identify the variables to be used.

E Read the problem carefully.

3 (4 marks) Solve these simultaneous equations.

$$x = -2 + 3y$$
$$y = x - 6$$

C S F

9780170454551

HW HOMEWORK

STARTUP ASSIGNMENT 10 (10)

WS WORKSHEET

PART A: BASIC SKILLS / 15 marks

1 Complete: $1\text{ m}^3 =$ ______ cm^3

2 Simplify $\frac{2a}{5} + \frac{a}{3}$. ______

3 Write one property of the diagonals of a rectangle. ______

______

4 Divide \$112 in the ratio 5 : 2. ______

5 Calculate, correct to 2 decimal places, the volume of a cylinder with diameter 6.6 cm and height 11 cm. ______

6 Convert a salary of \$47 800 to a weekly pay. (Answer to the nearest dollar.) ______

7 Find the value of y in the diagram below.

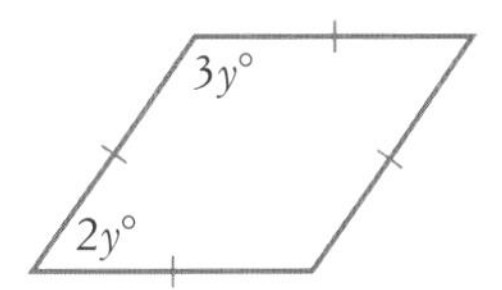

______

8 Find the gradient of a line parallel to $y = 3x - 6$. ______

9 If $y = \frac{k}{x}$, find k if $y = 7$ when $x = 10$.

______

10 Solve $3d - 4 = d + 10$. ______

11 A cap sells for \$22.54 after an 8% discount. What was its original price?

______

12 Calculate to the nearest \$100 the value of a \$29 000 car after 6 years if it depreciates at 16% per annum.

______

13 Find the value of r.

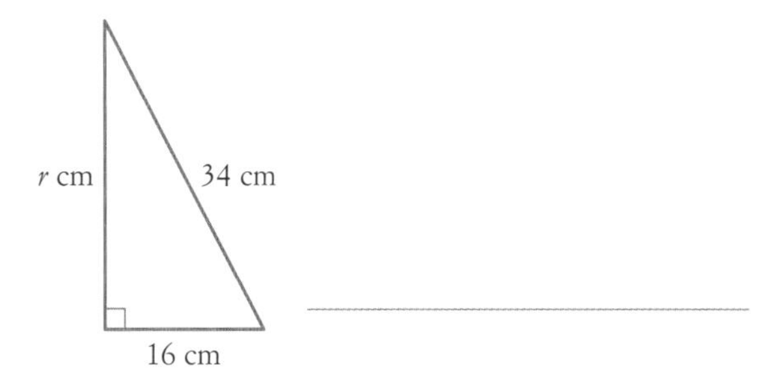

______

14 A cube has a volume of 13.824 m^3. Find its:

a length ______

b surface area. ______

PART B: PROBABILITY / 25 marks

15 If the chance that it does not rain on Saturday is 42%, what is the chance that it does rain?

______

16 What fraction of the numbers from 1 to 20 are:

a less than 8? ______

b square numbers? ______

c divisible by 4? ______

WS WORKSHEET

17 A ticket is drawn at random from a box containing 7 blue, 10 white, 4 red and 3 yellow tickets. Find the probability the ticket drawn is:

a red ______

b blue or white ______

c not red. ______

18 What is the probability a person chosen at random has a birthday in a month beginning with M?

______

19 Sort these events in order from most to least likely:

D rolling an odd number on a die

L buying your lunch at the canteen tomorrow

Y the next person visiting the class being in Year 9

P a letter arriving in the post for you today

I you watch YouTube tomorrow

S you are at school before 8:30 a.m. tomorrow

______

20 Evaluate:

a $1 - 0.45$ ______

b $1 - \frac{5}{6}$ ______

21 What does it mean if an event has a probability of 0? ______

22 Write as a decimal the value of a '50-50 chance'. ______

23 What are the possible outcomes for the result of a soccer match between the Reds and the Blues?

______

24 Henry, Irene, Jack, Kathy and Lisa wrote their names on separate cards. What is the probability that a card chosen at random has a boy's name?

______

25 There are 400 tickets in a raffle. If Johanna buys 8 tickets, find the decimal probability she wins first prize. ______

26 If the probability of having 2 boys in a 3-child family is $\frac{3}{8}$, what is the probability of not having 2 boys? ______

27 A tossed coin came up heads 18 times and tails 22 times. What percentage of tosses showed tails? ______

28 What fraction of a deck of cards are:

a diamonds? ______

b aces? ______

c jacks, queens or kings? ______

d even numbers? ______

29 What percentage (to one decimal place) of the alphabet are vowels?

______

PART C: CHALLENGE

Bonus / 3 marks

Thomas, Patrick, Adrian, Alexis and Chris sit together in a row for a group photo.

How many possible seating arrangements are there?

______

C S F

9780170454551

PROBABILITY CROSSWORD 

I'M ZINA. ALL THE ANSWERS TO THIS CROSSWORD ARE LISTED BELOW. YOU JUST HAVE TO WORK OUT WHERE THEY GO!

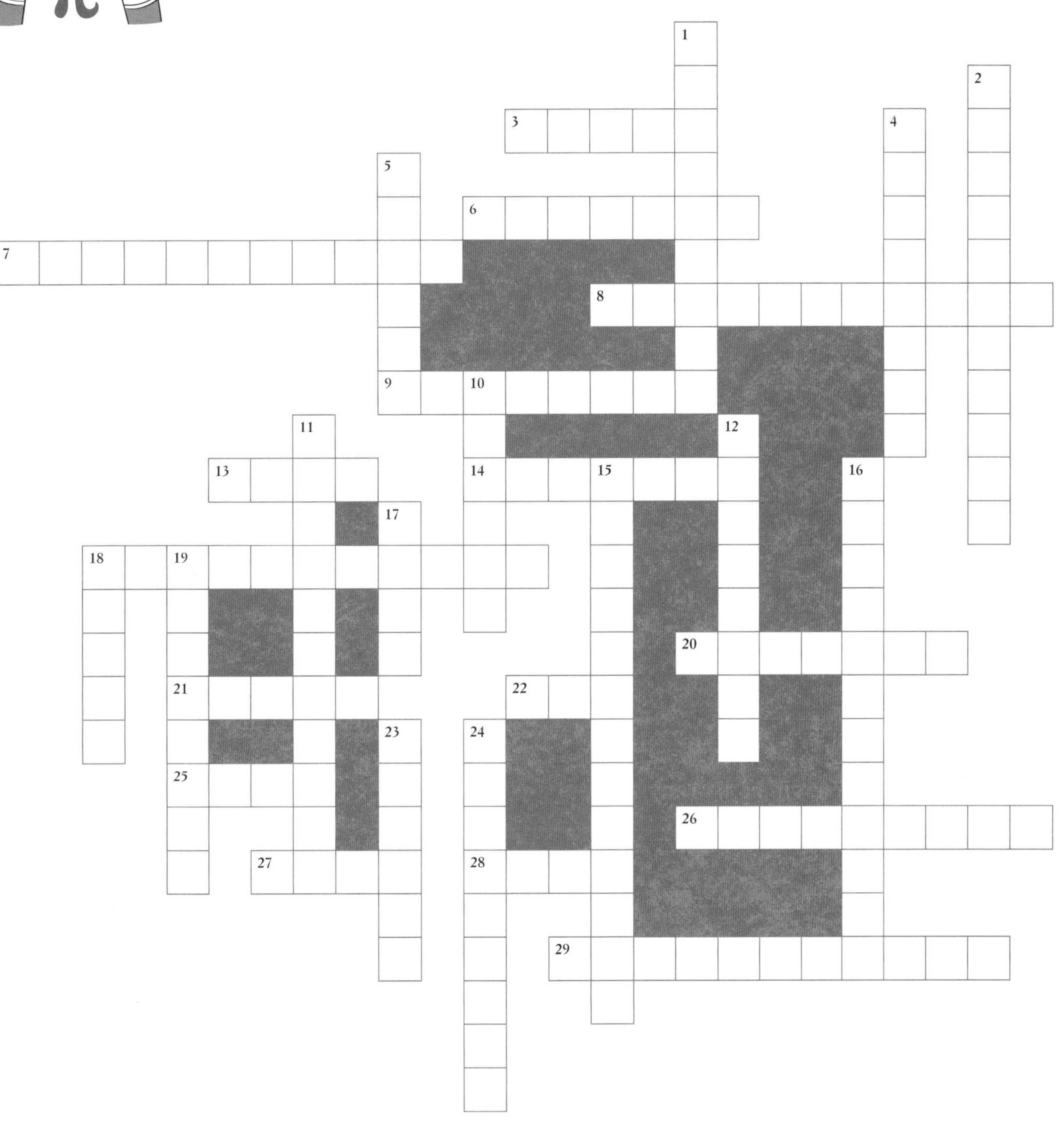

COMPLEMENTARY	COMPOUND	CONDITIONAL	DEPENDENT
DIAGRAM	DICE	DIE	EVENT
EXCLUSIVE	EXPECTED	EXPERIMENTAL	FREQUENCY
INDEPENDENT	LIST	MUTUALLY	OUTCOME
OVERLAPPING	PROBABILITY	RANDOM	RELATIVE
REPLACEMENT	SAMPLE	STEP	TABLE
THEORETICAL	TREE	TRIAL	TWO-WAY
VENN	WITHOUT		

WS WORKSHEET

10 TREE DIAGRAMS

TREE DIAGRAMS ARE OFTEN NEEDED FOR PROBABILITY PROBLEMS. MAKE SURE YOU PRACTISE DRAWING AND USING THEM.

1 **a** Explain what this tree diagram shows about two-child families.

First child	Second child	Outcomes
B	B	BB
	G	BG
G	B	GB
	G	GG

B = boy, G = girl

____

____

____

____

b Find the probability of a two-child family having:

i 2 girls ______________________

ii a boy followed by a girl ______________

iii one boy and one girl born in any order.

____

2 Piper sits a test in Science and in History.

a If she has an even chance of passing or failing each test, list all the possible outcomes on the tree diagram below.

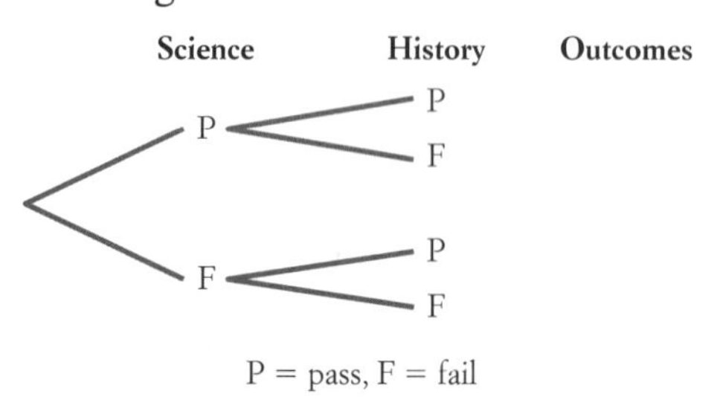

b Find the probability that Piper:

i passes both tests ______________

ii fails both tests. ______________

3 Justin wants to work out the chance of rain over a long weekend (3 days).

a If rain (R) or no rain ($\overline{R}$) on each day are equally likely, complete the tree diagram below and list the outcomes.

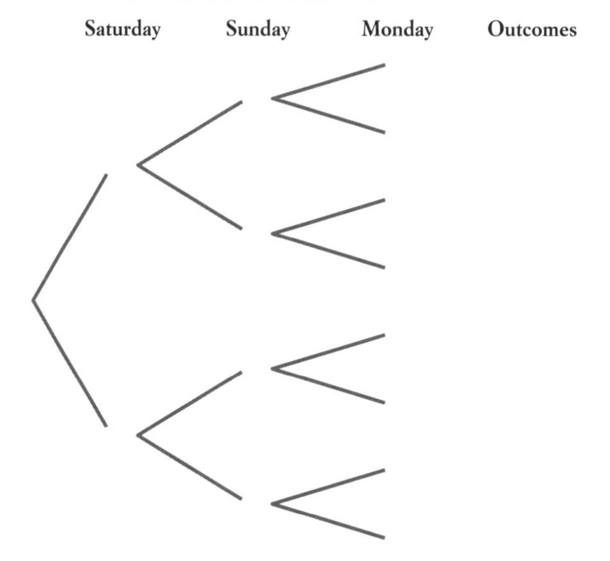

b Calculate the probability that, over the 3 days, it rains on exactly one of the days.

____

4 Adrian and Pete each shoot for goal on a basketball court. Each player has an equal chance of scoring or missing the goal. Complete the tree diagram below to help you find the probability that at least one of them scores.

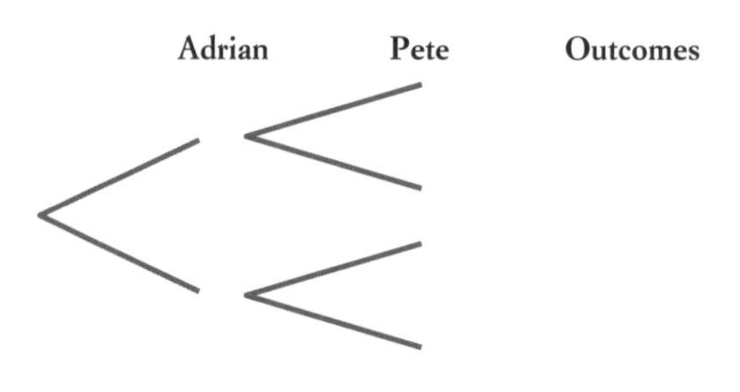

____

9780170454551

5 Anna rolls 2 dice and finds the difference between the 2 numbers rolled.

a Complete the table to show all the possible differences.

		1st die					
		1	**2**	**3**	**4**	**5**	**6**
2nd die	**1**	0					
	2						
	3						3
	4						
	5			2			
	6						

b Find the of probability of rolling a difference:

i of 3 ____________

ii of 0 ____________

iii that is even ____________

iv of more than 3. ____________

6 A 4-sided die (numbered 1, 2, 3, and 4) is rolled and a coin is tossed.

a Use a tree diagram or table to list all possible outcomes.

b Find the probability of obtaining:

i a 2 and a head ____________

ii an odd number and a tail ____________

7 A bag contains blue, red and white socks. Three socks are chosen from the bag. Use a tree diagram to find the probability that at least 2 of the socks are white. ____________

8 Which situation below can be illustrated by this tree diagram?

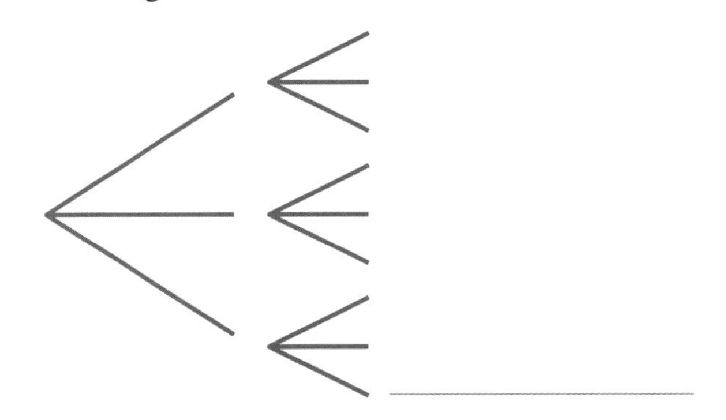

____

A Tossing a coin 3 times

B Selecting 2 balls from a bag of red, blue and green balls

C 3 students passing or failing an exam

9 Georgina, Megan and Allyson sit in a row for a photo. What is the probability that:

a Megan sits in the middle? ____________

b Allyson sits on either side? ____________

c Georgina sits further left than Megan?

____

WORKSHEET

WS WORKSHEET

10 TWO-WAY TABLES

A TWO-WAY TABLE IS A GREAT WAY OF DISPLAYING DATA ABOUT 2 TYPES OF CATEGORIES.

1 A sample of Year 7 students was surveyed on whether they owned a dog or cat.

	Have a dog	Do not have a dog
Have a cat	3	5
Do not have a cat	12	4

a How many students were surveyed?

____

b How many students have a cat?

____

c If a student is selected at random, what is the probability that he or she:

i has a cat and a dog? ____________

ii has neither a cat nor a dog? ____________

iii does not have a cat? ____________

2 This information was collected about the types of cars sold at a caryard.

	Automatic	Manual
Sedan	32	18
Hatchback	11	9

a How many cars were there? ____________

b How many cars were sedans? ____________

c If a car is selected at random, what is the probability it is:

i an automatic? ____________

ii a manual hatchback? ____________

d If a sedan is chosen at random, what is the probability it is a manual? ____________

3 People in a shopping centre were surveyed on their favourite leisure activity.

	Dance	Sport	TV
Female	3	12	7
Male	15	8	5

a How many people were surveyed?

____

b Which activity was the most popular?

____

c What is the probability that a person chosen at random from the survey:

i is female? ____________

ii said sport was their favourite activity?

____

iii is a male whose favourite activity is watching TV? ____________

iv is a female whose favourite activity is dance or sport? ____________

4 A hospital kept statistics about the birthweight of babies born in the last month.

	Less than 5 kg	**5 kg or more**
Female	25	7
Male	19	9

a How many girls were born? ______

b How many babies weighing 5 kg or more were born? ______

c What is the probability that a baby chosen at random from this hospital:

i is male? ______

ii weighs less than 5 kg? ______

iii is female and weighs 5 kg or more?

______

5 Year 9 students were surveyed on how they normally travel to school.

	Male	**Female**
Walk	0	0
Car	13	18
Bus	48	37
Cycle	7	2
Skateboard	5	0

a How many students were surveyed?

______

b Were there more males or more females surveyed?

______

c What is the probability that a Year 9 student selected at random:

i cycles to school? ______

ii is female and catches a bus to school?

______

iii uses a car or bus to get to school?

______

d Why do you think no-one walks to school?

______

10 PROBABILITY 1

Name:

Due date:

Parent's signature:

Part A	/ 8 marks
Part B	/ 8 marks
Part C	/ 8 marks
Part D	/ 8 marks
Total	/ 32 marks

PART A: MENTAL MATHS

Calculators are not allowed

1 (4 marks) Find x and y, giving reasons.

$x°$, $75°$, $y°$

2 (2 marks) Solve $\frac{5x-2}{7} = \frac{3x+2}{4}$.

3 (2 marks) Find the centre and radius of the circle with equation $x^2 + y^2 = 16$.

PART B: REVIEW

1 The probability of rain tomorrow is 0.75. What is the probability of no rain tomorrow?

2 A bag contains 2 red crayons, 3 yellow crayons and 5 blue crayons. A crayon is drawn at random from the bag and the colour recorded.

a How many colours are in the sample space?

b Are the colours equally likely? Explain your answer.

3 (3 marks) This two-way table shows the survey results on students' favourite pets. Complete the table.

	Dog	Cat	Others	Total
Girls	10	18		35
Boys		10		
Total	25		16	

4 James bought 5 tickets in a raffle in which 100 tickets were sold. What is the percentage probability that James wins 1st prize?

5 What is the expected frequency of tails if a coin is tossed 200 times?

C S F

9780170454551

PART C: PRACTICE

› Relative frequency
› Venn diagrams
› Two-way tables

1 (2 marks) Lexie spun a spinner with 5 colours many times and recorded the results. Complete the table.

Colour	Frequency	Relative frequency
Yellow	20	$\frac{1}{5}$
Red	32	
Black	24	$\frac{6}{25}$
Blue	16	$\frac{4}{25}$
Green	8	

2 (2 marks) This Venn diagram shows the results of a survey on what types of pizza students like.

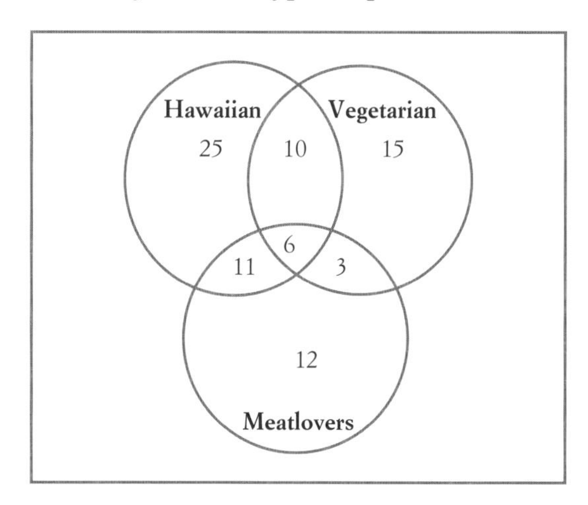

How many students like:

a both Vegetarian and Meatlovers?

b Hawaiian or Vegetarian but not both?

3 (4 marks) For 110 Year 10 students, 45 study Music, 56 study Art, and 24 do not study Music or Art. Show this information on a Venn diagram.

PART D: NUMERACY AND LITERACY

1 (2 marks) Complete:

$$P(E) = \frac{\text{number of favourable \_\_\_\_\_\_}}{\text{total number of \_\_\_\_\_\_}}$$

2 In the above formula, what does E stand for?

3 What 2-word phrase describes the expected number of times an event will occur over repeated trials, such as the number of days it will rain in December?

4 (4 marks) A group of students were asked about their favourite types of movies.

	Girls	Boys
Comedy	20	22
Horror	10	11
Action	14	34
Romance	30	9

a How many students were surveyed?

b What is the probability of randomly selecting a student who:

i prefers comedy?

ii is a boy who likes horror movies?

iii is a girl who doesn't like action movies?

C S F

HW HOMEWORK

10 PROBABILITY 2

THIS PROBABILITY TOPIC IS GETTING QUITE HARD, SO YOU'LL NEED TO PRACTISE AND SHARPEN YOUR SKILLS.

Name:

Due date:

Parent's signature:

Part	Marks
Part A	/ 8 marks
Part B	/ 8 marks
Part C	/ 8 marks
Part D	/ 8 marks
Total	/ 32 marks

PART A: MENTAL MATHS

Calculators are not allowed

1 Convert 63.715° to degrees and minutes, to the nearest minute.

2 Simplify $4a^{20}b^{35} \div 12a^4b^5$.

3 Find:

a the scale factor for these similar figures

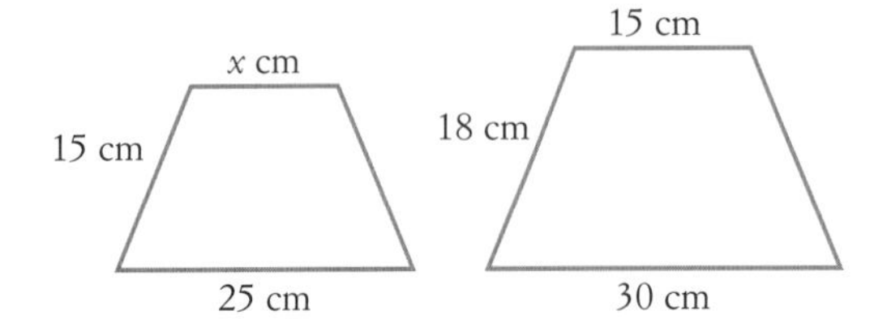

b the value of x.

4 Find the shaded area in terms of π.

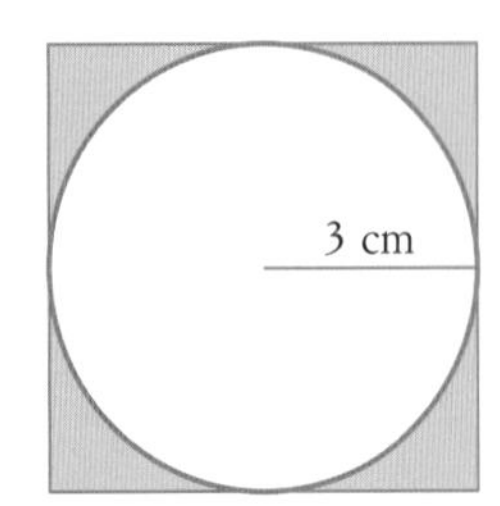

5 (2 marks) Solve $2(3y + 1) + 5 = 4(4y + 6)$.

6 Write 25 365 000 in scientific notation.

PART B: REVIEW

1 From a group of 945 people, what is the expected frequency of randomly selecting a person who was born on a Tuesday?

2 (5 marks) This Venn diagram shows whether 50 people liked cake or lollies.

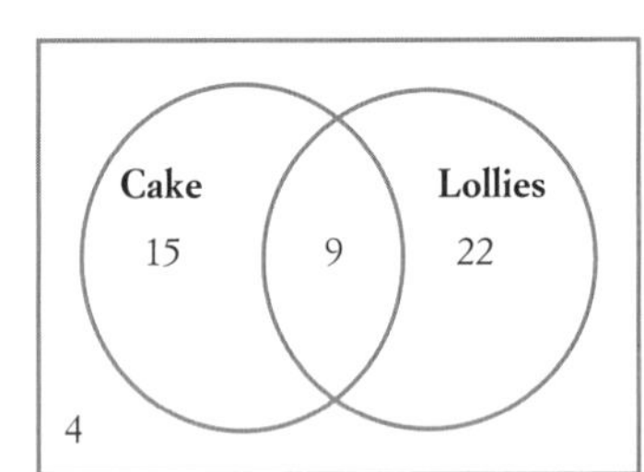

a How many people liked lollies?

b How many people did not like cake or lollies?

c Complete this two-way table.

	Cake	Not cake
Lollies		
Not lollies		
Total		

C S F

9780170454551

3 (2 marks) 2 coins are tossed together 40 times and the number of heads counted each time.

Number of heads	Frequency
0	8
1	23
2	9
Total	40

What is the relative frequency of tossing:

a 2 heads? ______

b 2 tails? ______

PART C: PRACTICE

› Tree diagrams
› Dependent and independent events
› Conditional probability

1 (4 marks) 3 coins are tossed.

a Draw the tree diagram to show all possible arrangements of heads (H) and tails (T).

b How many outcomes are there in the sample space? ______

c Find the probability of tossing at least one head. ______

C
S
F

2 (3 marks) A bag contains 6 blue socks and 4 green socks. 2 socks are chosen randomly without replacement. Find the probability of choosing:

a a blue sock first

______

b a green sock second after a green sock was chosen first

______

3 2 dice are rolled. What is the probability of rolling a 3 on the second die, given that a 3 was rolled on the first die?

______

PART D: NUMERACY AND LITERACY

1 What type of diagram has circles (usually overlapping) for grouping items into categories?

______

2 How many different outcomes are possible when 2 dice are rolled? ______

3 (6 marks) 2 buttons are drawn without replacement from a bag containing 3 yellow and 2 blue buttons. Draw a tree diagram for the 20 possible outcomes and use it to find the probability that:

a both buttons are yellow ______

b the second button is blue ______

c the second button is yellow, given that the first button is yellow

______

d both buttons are blue, given that both buttons are the same colour

______

HW HOMEWORK

11 STARTUP ASSIGNMENT 11

THIS ASSIGNMENT WILL HELP US TACKLE THE GEOMETRY TOPIC.

PART A: BASIC SKILLS

/ 15 marks

1 Calculate, correct to 2 decimal places:

a 5.7 cos 42°31′ ______

b the value of a $27 000 car after 2 years if it depreciates at 11% p.a. ______

c the volume of a cylinder with radius 6 m and height 6 m ______

2 Describe the **mode** of a list of scores.

______

3 Find the scale on a map if 4 cm represents 200 m. ______

4 Find the average of b, $b + 10$ and $b - 1$.

______

5 What is the bearing of P from O?

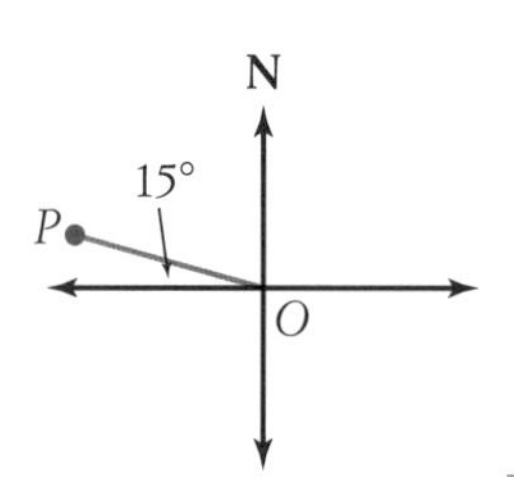

______

6 Solve $3(x + 7) = 5x + 14$. ______

7 Find the value of x in the diagram below.

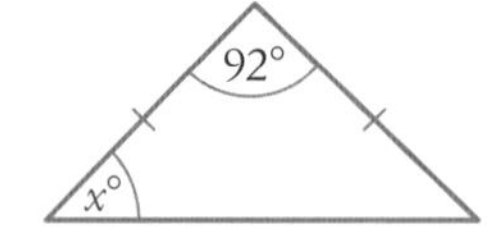

______

8 Maddison walks at 6 km/h for $1\frac{3}{4}$ hours. How far does she walk? ______

9 Find the surface area of a cube with side length 4.5 cm. ______

10 What size angle does the minute hand of a clock turn in 4 minutes? ______

11 For $X(-2, 3)$ and $Y(-10, -12)$ on the number plane, find the length of XY. ______

12 Calculate Yasmin's pay if she earns $12.35 per hour for 8 hours plus $2\frac{1}{2}$ hours at time-and-a-half. ______

13 What is the gradient of $4x - 8y + 24 = 0$?

______

PART B: GEOMETRY

/ 25 marks

14 Measure the size of $\angle JOK$ below.

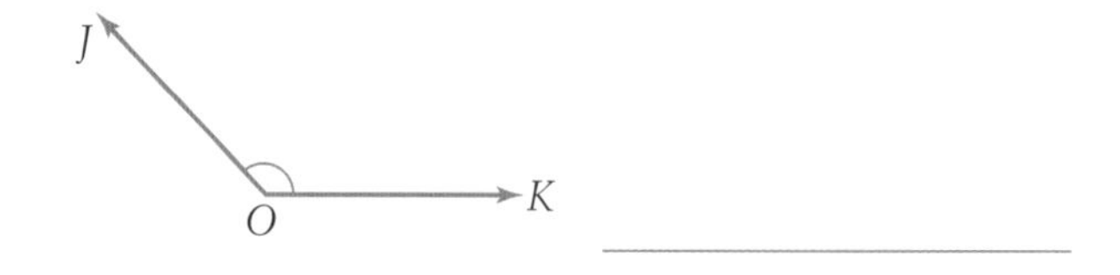

______

15 Draw a rhombus and mark its axes of symmetry.

16 Which angle is vertically opposite $\angle CEB$ below?

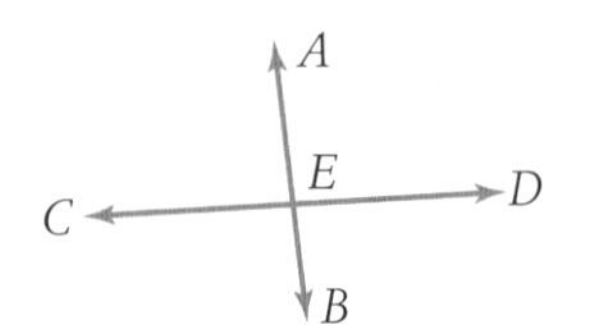

______

17 What is the angle sum of a quadrilateral?

______

18 What type of triangle has all angles 60°?

______

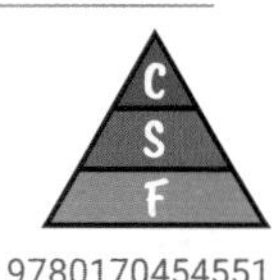

9780170454551

19 A map's scale indicates that 1 cm represents 50 km.

a What distance does 5.5 cm on the map represent? ______

b 2 towns are 325 km apart. What is their scaled distance on the map?

______

c Write the scale as a simplified ratio.

1 : ______

20 Draw an obtuse-angled triangle.

21 Find the value of each variable.

a

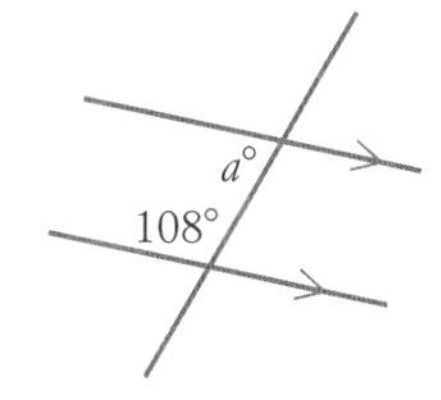

$a =$ ______

b

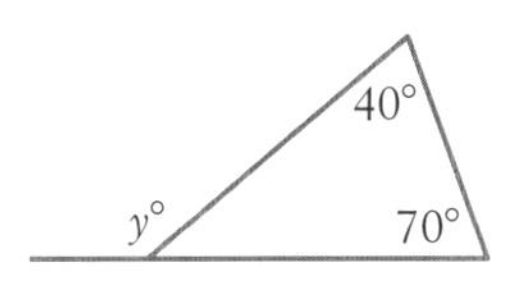

$y =$ ______

c

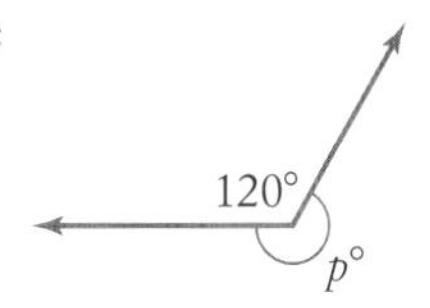

$p =$ ______

d

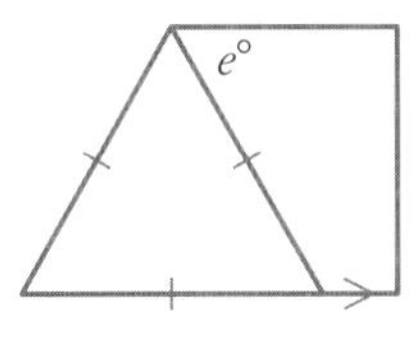

$e =$ ______

e

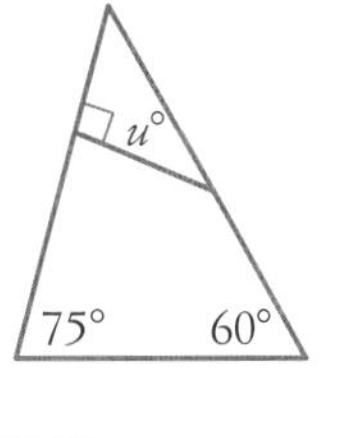

$u =$ ______

f

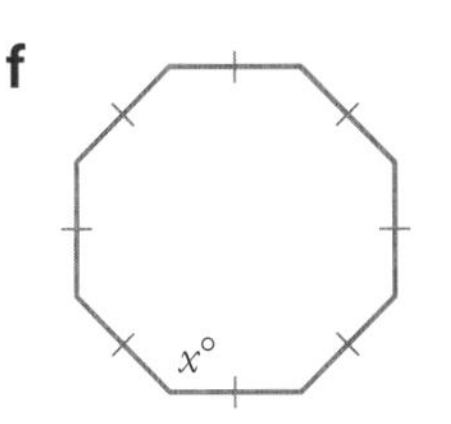

$x =$ ______

22 How many sides has an octagon? ______

23 True or false? The diagonals of a parallelogram:

a are equal ______

b bisect each other ______

c cross at right angles ______

24 If an isosceles triangle has a 130° angle, what are the sizes of the other 2 angles?

______

25 **a** Which 2 triangles are congruent?

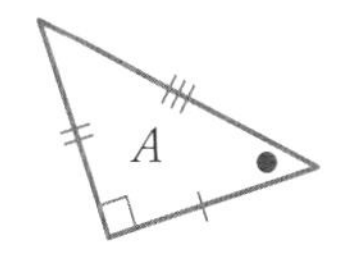

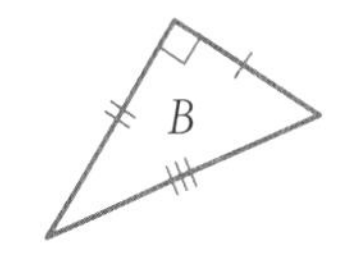

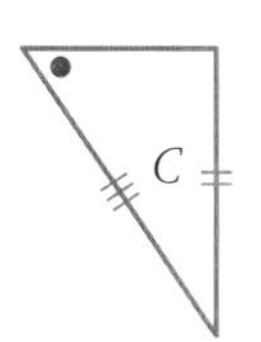

______

b Which congruence test can be used for these triangles? ______

26 **a** If $\triangle ABC \,|||\, \triangle JLK$, which angle matches $\angle A$?

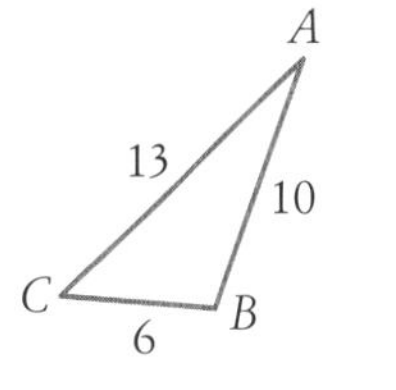

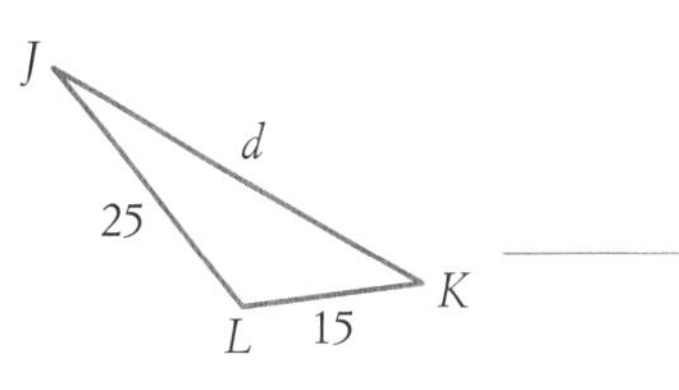

______

b Find the value of d. ______

PART C: CHALLENGE

Bonus / 3 marks

Place 5 crosses on this grid so that no 2 crosses are in the same row, column or diagonal.

C S F

WS WORKSHEET

11 PROVING PROPERTIES OF QUADRILATERALS

WE'RE USING CONGRUENT TRIANGLES TO PROVE PROPERTIES OF SPECIAL QUADRILATERALS.

WS WORKSHEET

1 The kite

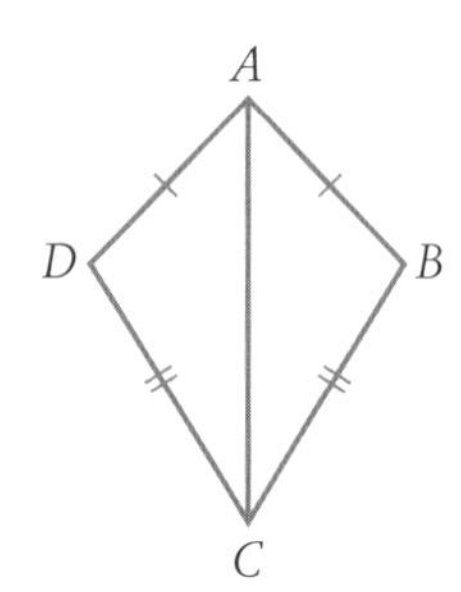

a If $ABCD$ is a kite, which congruence test proves that $\triangle ADC \equiv \triangle ABC$?

____

b Mark all pairs of matching angles on kite $ABCD$.

c Complete: One of the diagonals of a kite is an ____________ of symmetry which ____________ two angles of the kite.

d Draw diagonal DB on the kite to cross AC at E.

e Why is it true that $\triangle ADE = \triangle ABE$?

____

f Which congruence test proves that $\triangle ADE \equiv \triangle ABE$? ________________

g Which angle is equal to $\angle AED$?

____

h What is the size of $\angle AED$?

____

i Complete: The diagonals of a kite cross at

______________________________.

2 The parallelogram

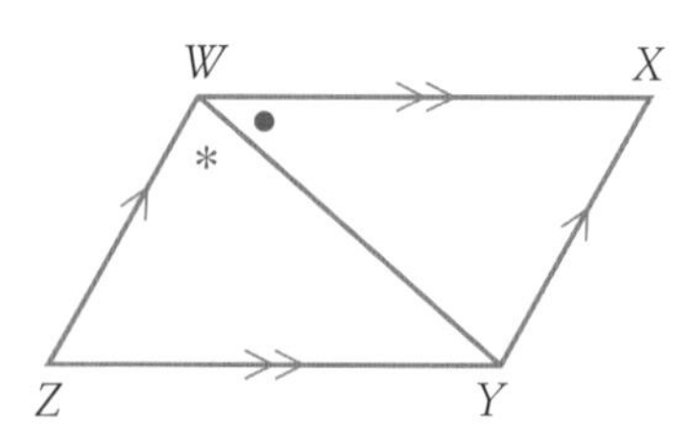

a In this parallelogram, which congruence test proves that $\triangle ZWY \equiv \triangle XYW$?

____

b Which angle is equal to $\angle WZY$?

____

c Mark all pairs of matching angles on $WXYZ$.

d Complete: In a parallelogram, opposite angles are ____________ and opposite sides are ____________.

e Draw diagonal ZX to cross WY at O.

f Why is it true that $\angle WOZ = \angle XOY$?

____

g Which congruence test proves that $\triangle WOZ \equiv \triangle XOY$? ________________

h Mark all pairs of matching sides on $WXYZ$.

i Complete: The diagonals of a parallelogram ________________ each other.

9780170454551

3 The rhombus

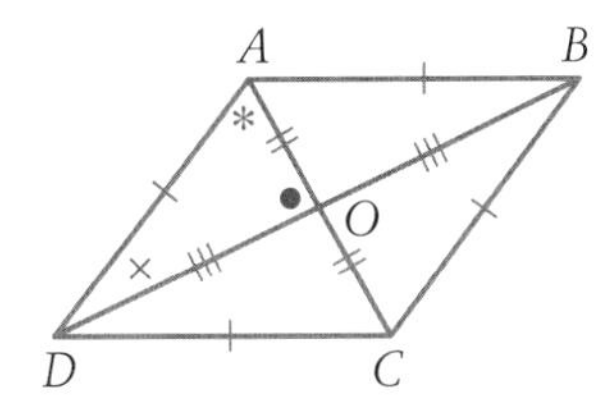

a In this rhombus, which test proves that $\triangle AOD \equiv \triangle AOB \equiv \triangle COB \equiv \triangle COD$?

____

b Mark all pairs of matching angles on $ABCD$.

c What are the sizes of the 4 angles at the centre of the rhombus? _______________

d Complete: In a rhombus, the diagonals bisect at _______________ and also _______________ the 4 angles of the rhombus.

4 The rectangle

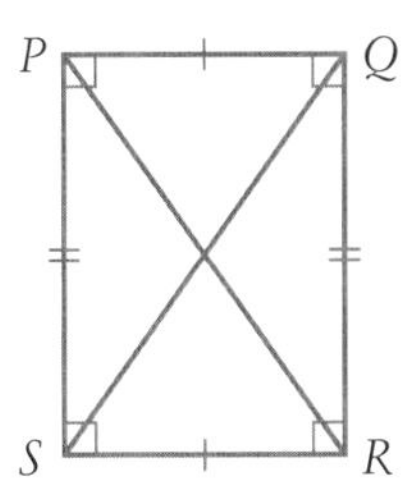

a If $PQRS$ is a rectangle, which congruence test proves that $\triangle PSR \equiv \triangle QRS$?

____

b Which side of $\triangle QRS$ is equal to PR?

____

c Complete: The diagonals of a rectangle are _______________.

5 The square

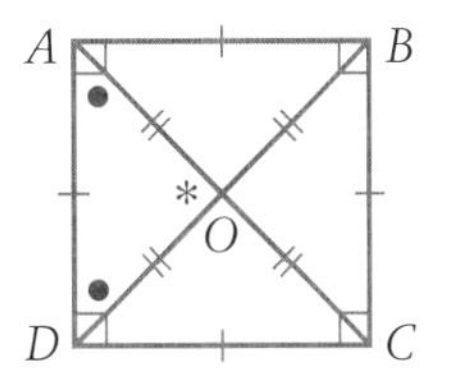

a A square is a special type of rectangle and parallelogram. What does this mean about its diagonals?

____

b For square $ABCD$, which test proves that $\triangle AOD \equiv \triangle AOB \equiv \triangle BOC \equiv \triangle DOC$?

____

c Why is it true that $\angle OAD = \angle ODA$?

____

d Mark all pairs of matching angles in square $ABCD$.

e What is the size of $\angle OAD$ and its matching angles? _______________

f What are the sizes of the 4 angles at the centre of the square? _______________

g Complete: The diagonals of a square are _______________ and bisect each other at _______________ _______________. The diagonals also _______________ the right angles of the square.

11 GEOMETRY CROSSWORD

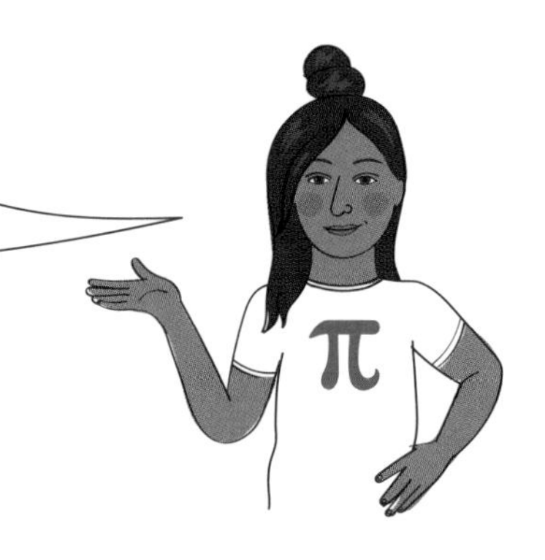

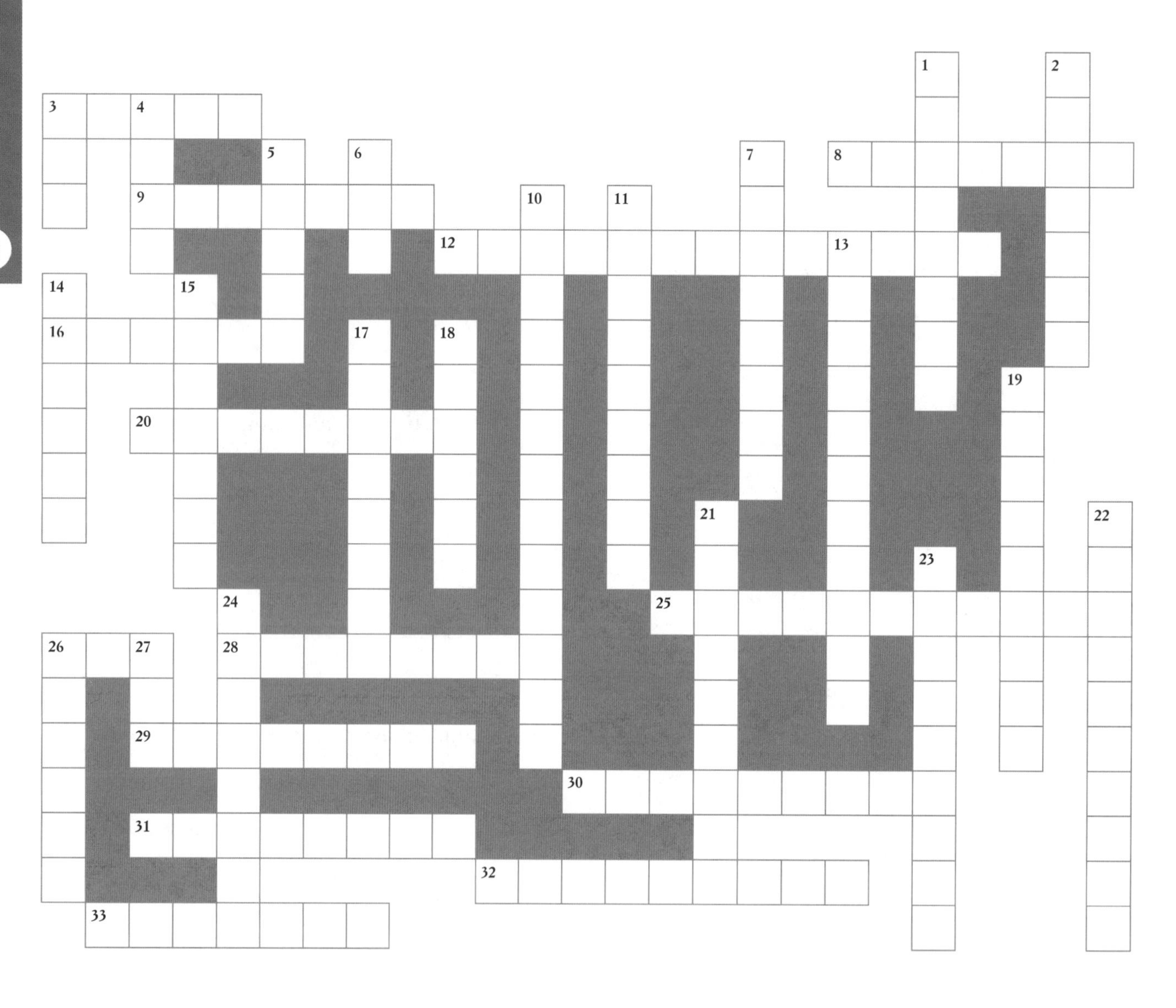

9780170454551

Clues across

3 Similar figures have sides whose lengths are in the same r__________

8 9-sided polygon

9 Same shape, different size

12 The name for a 4-sided polygon

16 To cut in half

20 In congruent figures, __________ sides are equal

25 The process of making a shape bigger

26 Side-side-side (abbreviation)

28 'Outside' angle of a triangle

29 A kite has one axis of ________________

30 12-sided shape

31 The opposite of 28 across

32 The opposite of 25 across

33 Any 2D shape with straight sides

PUZZLE SHEET

Clues down

1 5-sided military headquarters

2 A 'pushed-over' square

3 Right angle-hypotenuse-side (abbreviation)

4 There are 4 of these for proving congruent triangles

5 90° angle

6 Angle-angle-side (abbreviation)

7 Interval joining one vertex to an opposite vertex

10 Quadrilateral whose opposite angles are equal

11 Quadrilateral with one pair of parallel sides

13 Each angle in this triangle is 60°

14 Angle between 90° and 180°

15 10-sided polygon

17 An exterior angle of this equals the sum of the 2 interior opposite angles

18 The angle sum of a triangle is one hundred and __________ degrees

19 Triangle with 2 equal angles

21 Geometrical word for identical

22 This type of symmetry requires spinning a figure

23 Quadrilateral of right angles

24 Branch of mathematics that angles and shapes belong to

26 Diagonals bisect the angles of this shape at 45°

27 One of the congruent triangle tests (abbreviation)

11 CONGRUENT FIGURES

Name:

Due date:

Parent's signature:

Part A	/ 8 marks
Part B	/ 8 marks
Part C	/ 8 marks
Part D	/ 8 marks
Total	/ 32 marks

PART A: MENTAL MATHS

Calculators not allowed

1 Find x as a surd.

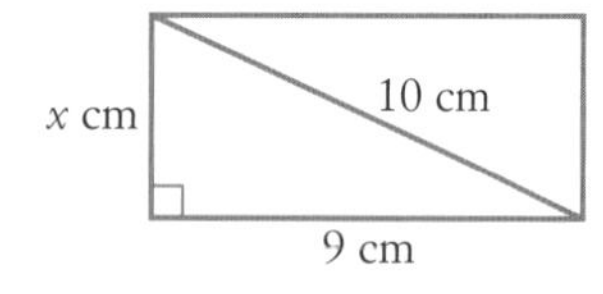

2 Calculate the simple interest for \$1200 invested at 6% p.a. for 2 years.

3 (2 marks) Simplify 72 : 96 : 36.

4 (2 marks) Factorise $x^2 + 4x - 21$.

5 A discount of \$105 represents 5% of the marked price. What is the marked price?

6 Find the size of an exterior angle in an equilateral triangle.

PART B: REVIEW

1 **a** (4 marks) Which congruence test proves that these triangles are congruent?

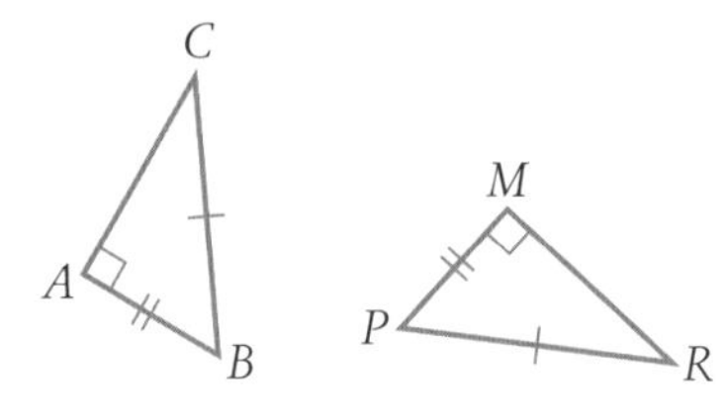

b Which side is equal to CB?

c Which angle is equal to $\angle C$?

d Complete: $\triangle CBA \equiv \triangle$ ______

2 Which test proves that each pair of triangles are congruent?

a

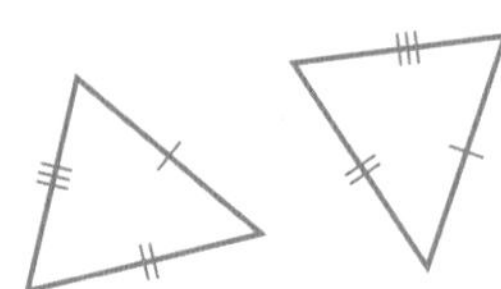

C S F

9780170454551

b

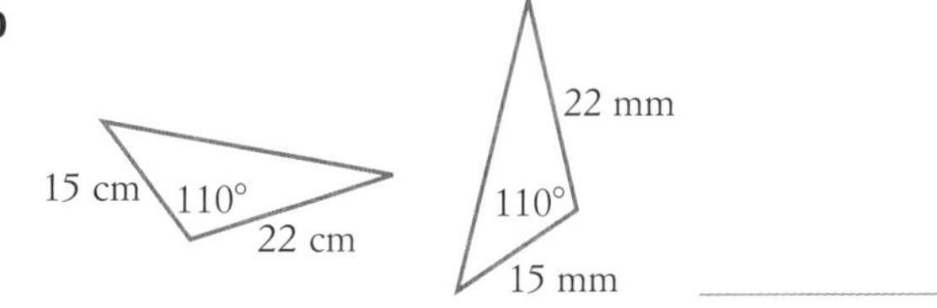

3 For these congruent triangles, find d and y.

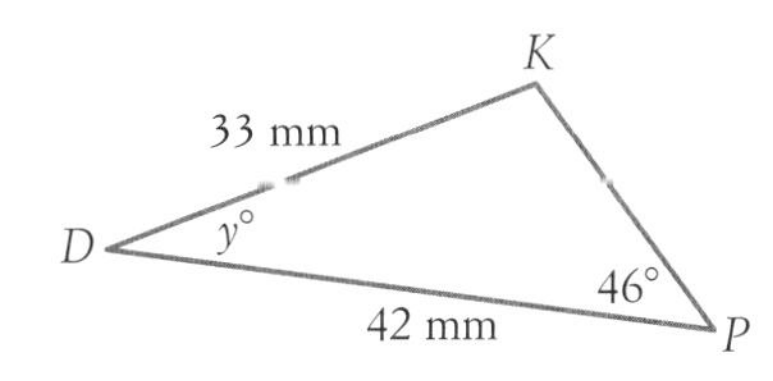

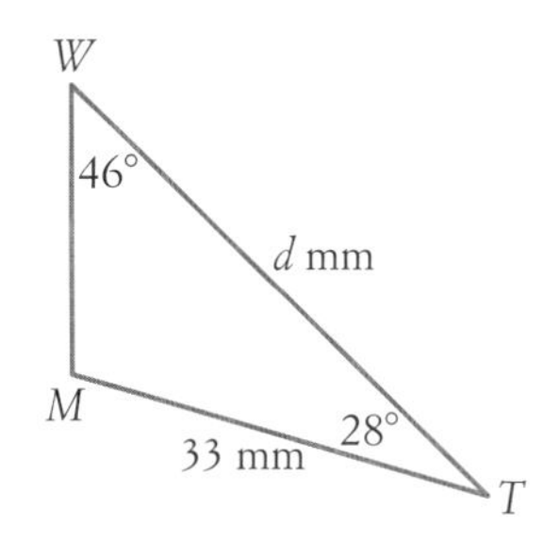

PART C: PRACTICE

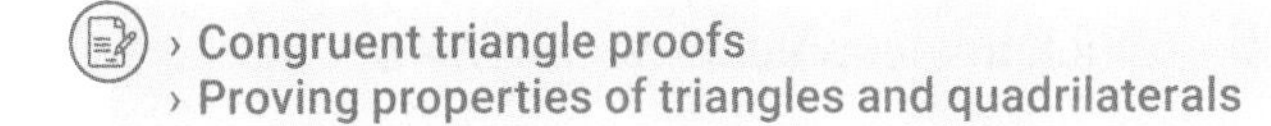

$AB = AC$ in the triangle and X is the midpoint of BC.

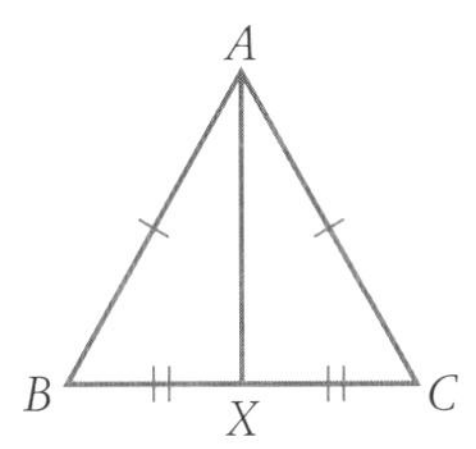

1 (4 marks) Prove that $\triangle ABX \equiv \triangle ACX$.

2 Complete: $\angle AXB = \angle$ ______ = ______°.

3 Which angle matches with $\angle B$? ______

4 What does this prove about an isosceles triangle?

PART D: NUMERACY AND LITERACY

$DCBA$ is a rectangle.

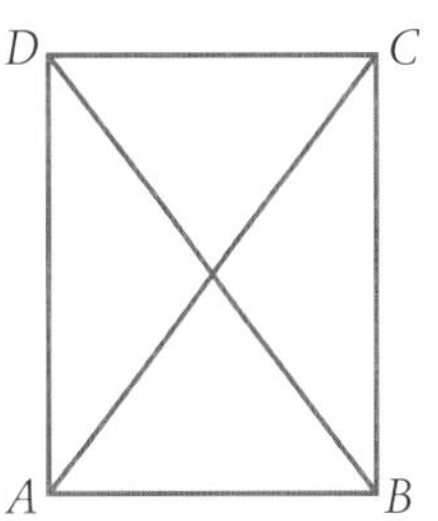

1 (4 marks) Prove that $\triangle DAB \equiv \triangle CBA$.

2 Which angle matches with $\angle ADB$? ______

3 Which side matches with DB? ______

4 What does this prove about the diagonals of a rectangle?

5 In the congruence test SAS, what is the special name given to A?

HW HOMEWORK

C S F

11 SIMILAR FIGURES

Name:

Due date:

Parent's signature:

Part	Marks
Part A	/ 8 marks
Part B	/ 8 marks
Part C	/ 8 marks
Part D	/ 8 marks
Total	/ 32 marks

PART A: MENTAL MATHS

Calculators not allowed

1 Simplify each ratio.

a 4 hours : 1 day

b \$2.80 : 20c

2 Evaluate $2\frac{4}{5} \div 1\frac{1}{4}$.

3 Simplify $-24a^2b^4 \div 4ab$.

4 Find x and y.

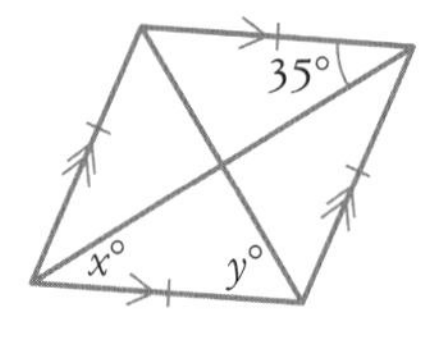

5 For this data set, find:

37 36 40 41 38 38 42 36 39

a the mode

b the median

PART B: REVIEW

1 (6 marks) In $\triangle ABC$, $AB = AC$ and $AD \perp BC$.

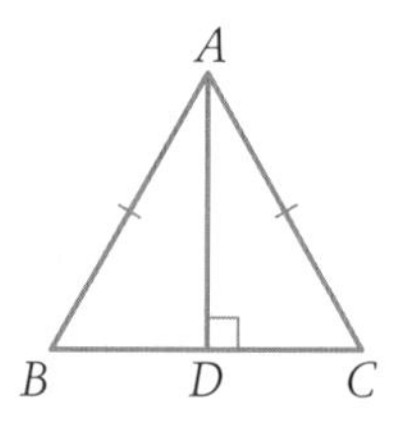

a Which congruence test can be used to prove that $\triangle ABD \equiv \triangle ACD$?

b Which angle matches with $\angle BAD$?

c Prove that $\triangle ABD \equiv \triangle ACD$ and hence show that AD bisects BC.

2 **a** Which similarity test proves that these triangles are similar?

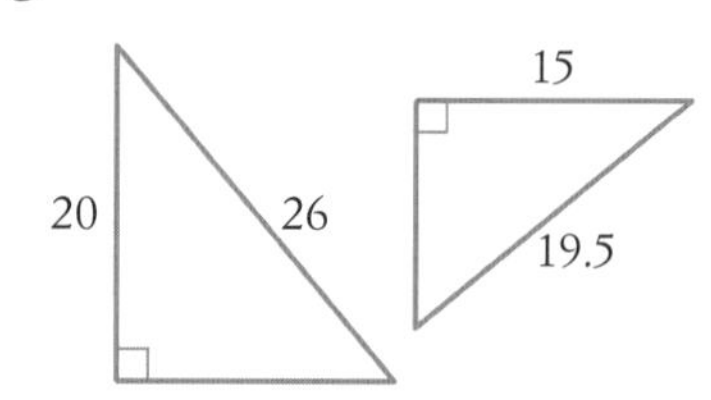

C S F

9780170454551

b What is the scale factor if the left triangle is the original? ______

PART C: PRACTICE

› Finding unknown lengths in similar figures
› Tests for similar triangles

1 (6 marks) Find the value of each variable for each pair of similar figures.

a

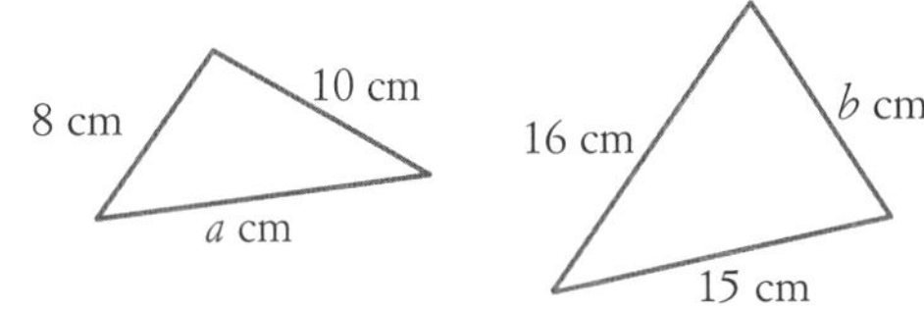

______

______

______

b

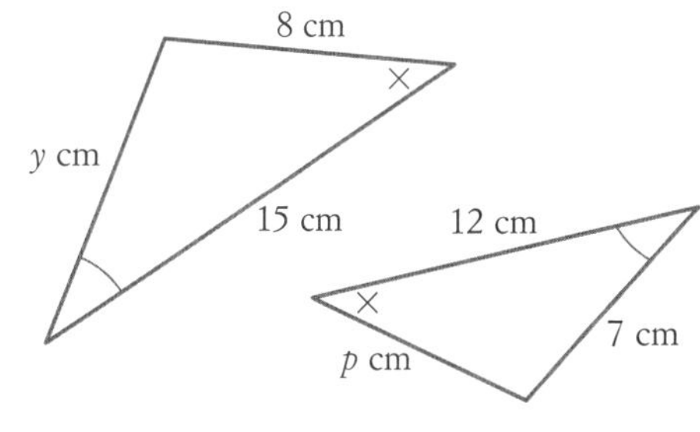

______

______

______

c

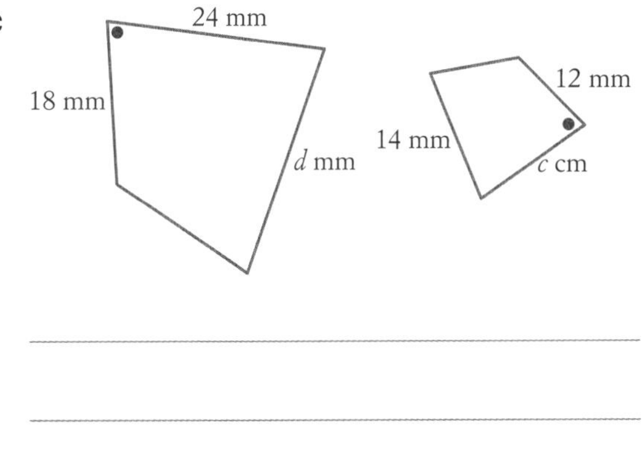

______

______

______

2 Complete: The similarity test 'SSS' means that the 3 sides of one ______ are ______ to the 3 sides of the other ______.

PART D: NUMERACY AND LITERACY

1 Complete:

a For an enlargement, the scale factor is greater than ______.

b For similar figures, matching sides are in the same ______ and matching angles are ______.

2 **a** (5 marks) In the diagram below, why is $\angle E = \angle BCA$?

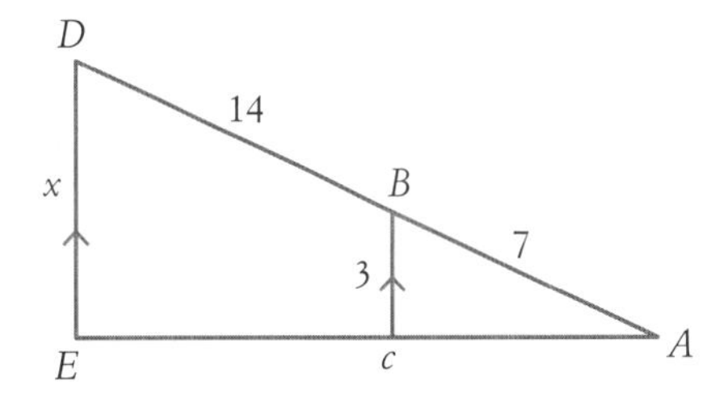

b There are 2 similar triangles in this figure. Complete: $\triangle ABC \mid\mid\mid \triangle$ ______.

c Which similarity test proves that these triangles are similar? ______

d Find the value of x.

______

______

HW HOMEWORK

11 GEOMETRICAL PROOFS ORDER ACTIVITY

Each card shows part of a geometry proof. Make an enlarged copy of the cards and cut them out.

Place each card, in order, under the appropriate problem to show the steps of each proof.

1

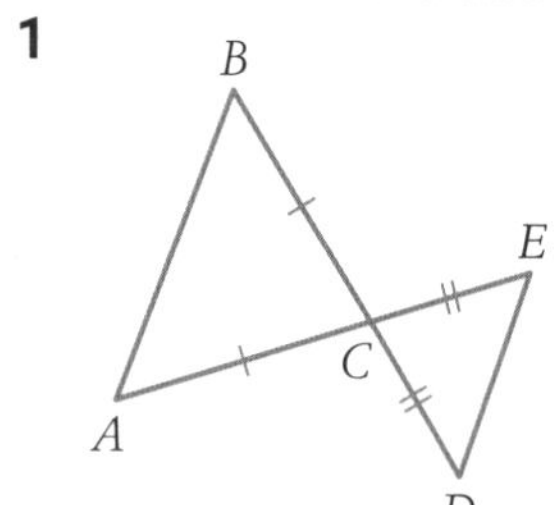

$BC = AC$, and $CE = CD$. Prove that $BA \parallel ED$.

2

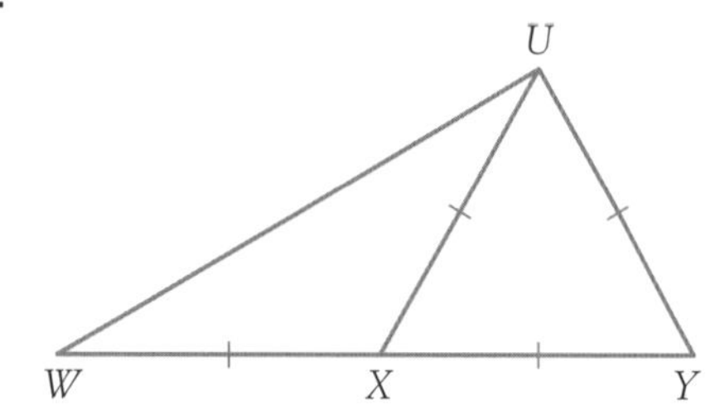

$\triangle UYX$ is equilateral and $WX = UX$.

Prove that $\angle WUY$ is a right angle.

3

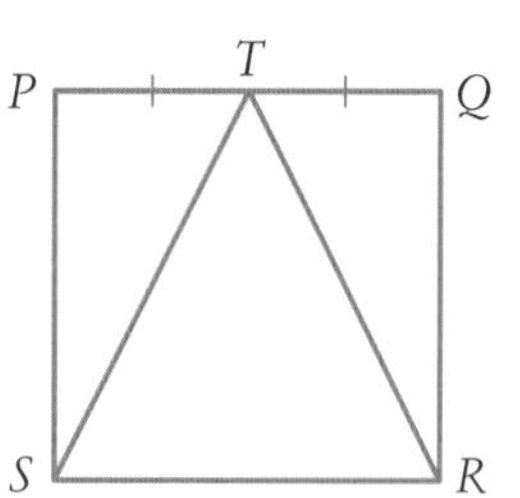

$PQRS$ is a square. T is the midpoint of PQ.

Use congruent triangles to prove that $\triangle TRS$ is isosceles.

Card
4 $\angle WUY = 90°$
5 $\therefore \angle SPT \equiv \angle RQT$ (SAS)
6 $\angle P = \angle Q = 90°$ (angles in a square)
7 $\therefore \angle WUY = 30° + 60°$
8 $\therefore \angle WXU = 180° - 60°$ (angles in a straight line)
9 $\angle WXU = 120°$
10 $\therefore BA \parallel ED$ (alternate angles are equal)
11 In $\triangle SPT$ and $\triangle RQT$
12 $\angle A = \angle E$
13 Let $\angle BCA = x$.
14 $\therefore x = \angle BCA = \angle ECD$ (vertically opposite)
15 $\angle WUX = 30°$
16 $PS = QR$ (sides of a square)
17 $\therefore TS = TR$ (matching sides in congruent triangles)
18 $\therefore \angle WUX = \dfrac{180° - 120°}{2}$ (angle sum of isosceles $\triangle XWU$)
19 $\therefore \angle WUY$ is a right angle
20 $PT = TQ$ (T is midpoint of PQ)
21 $\therefore \triangle TRS$ is isosceles
22 $\angle E = \dfrac{180° - x}{2}$ (angle sum of isosceles $\triangle EDC$)
23 $\angle XUY = \angle UXY = 60°$ (angles in equilateral $\triangle UYX$)
24 $\therefore \angle A = \dfrac{180° - x}{2}$ (angle sum of isosceles $\triangle BCA$)

9780170454551

STARTUP ASSIGNMENT 12

PART A: BASIC SKILLS

/ 15 marks

1 Evaluate $2x^2 + x - 6$ if $x = -1$.

2 What amount must be invested at 7.5% p.a. for 2 years to earn \$2700 simple interest?

3 Find the size of an exterior angle in a regular octagon.

4 How many hours and minutes will it take a truck moving at 84 km/h to travel 224 km?

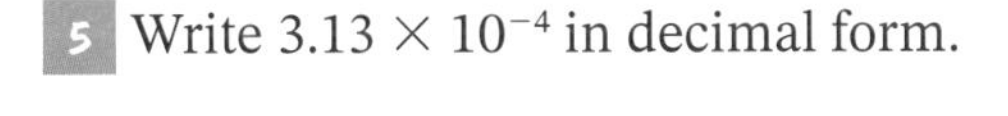

5 Write 3.13×10^{-4} in decimal form.

6 Find the interquartile range of these values:

3, 8, 5, 12, 7, 2.

7 Which quadrilateral has 4 equal angles?

8 What is the probability that the next 2 children born are both girls?

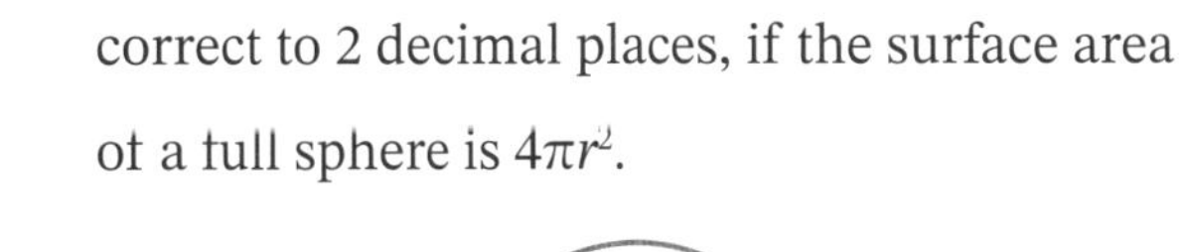

9 Calculate the surface area of this hemisphere, correct to 2 decimal places, if the surface area of a full sphere is $4\pi r^2$.

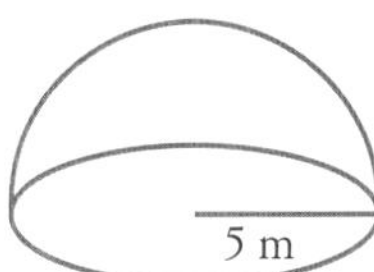

10 Graph the line $2x - 3y = 6$ on a number plane.

11 If O is the centre of the circle, which congruence test proves that $\triangle AOB \equiv \triangle DOC$?

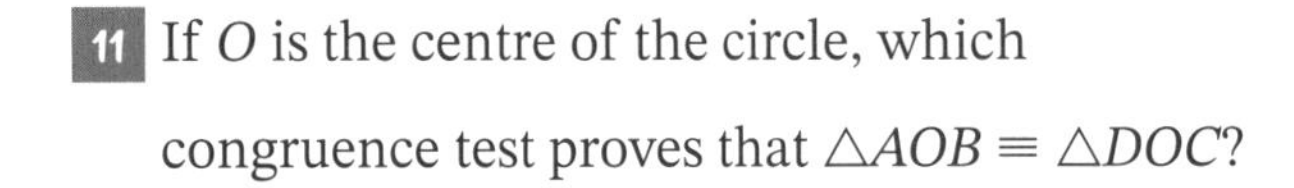

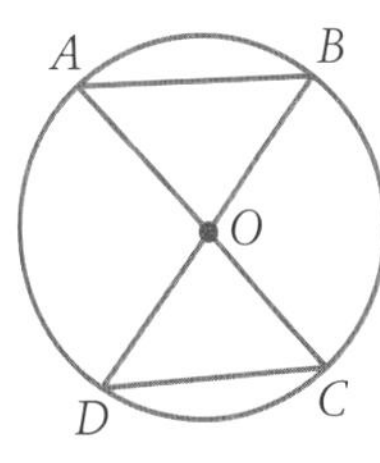

12 Make a the subject of $v^2 = u^2 + 2as$.

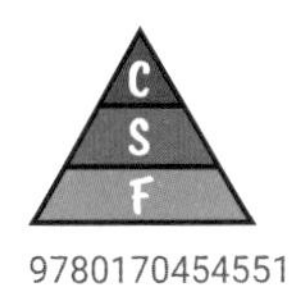

9780170454551

WS WORKSHEET

13 After an 11% discount, a jacket sells for \$40.05. What was its original price?

14

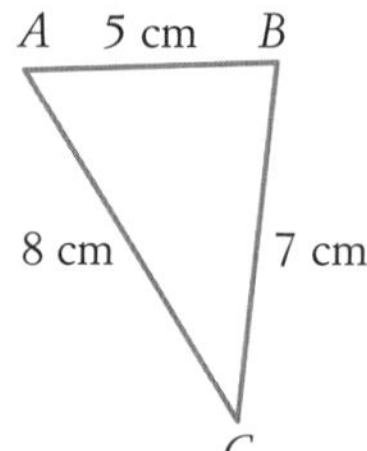

$\cos C = \dfrac{a^2 + b^2 - c^2}{2ab}$

$\text{Area} = \dfrac{1}{2} ab \sin C$

a Find the size of $\angle C$ to the nearest minute.

b Hence calculate the area of $\triangle ABC$ to one decimal place.

PART B: ALGEBRA AND GRAPHS

/ 25 marks

15 List the factors of 15.

16 Write the equation of a circle with centre (0, 0) and radius 4.

17 A graph has equation $y = 2x^2 + 8x - 10$.

a What type of graph is it ?

b Find its y-intercept.

c Find its x-intercepts.

18 Expand and simplify:

a $4(3m + 5) - 2(4 - m)$

b $(x - 3)(2x^2 + 5x)$

19 What is the quotient and remainder when 147 is divided by 6?

20 Find $4875 \div 25$ by long division.

21 If $y = (x - 5)(x + 1)(x - 2)$, find y when:

a $x = 3$

b $x = -1$

c $x = 0$

22 Factorise:

a $3b^2 + b - 14$

b $2y^3 + 14y^2 + 24y$

C S F

9780170454551

WORKSHEET

23 For $y = 2x^3 - 7x + x + 6$, write:

a the highest power of x ____________

b the coefficient of x^3 ____________

c the constant term ____________

24 Complete the square:

$x^2 + 10x + \_\_\_\_ = (x + \_\_\_\_)^2$

25 Sketch the graph of:

a $y = -x^3$

b $y = x^2 + 5$

c $y = 2 - x$

26 Solve:

a $x^2 + x - 56 = 0$

____

____

____

b $9p^2 - 48p + 64 = 0$

____

____

____

____

27 Solve the simultaneous equations:

$2x + 5y = 29$

$y = x^2 + 1$

____

____

____

____

____

____

____

____

PART C: CHALLENGE

Bonus / 3 marks

A group of 150 Year 10 students wanted to text the same message to each other by SMS.

The first student messaged 3 other students, each of whom messaged another 3, and so on, until all 150 students had been contacted.

If nobody received the message twice, how many students did not need to send the message?

C S F

12 SPECIAL PRODUCTS

PERFECT SQUARES AND DIFFERENCE OF 2 SQUARES ARE COVERED HERE.

WS WORKSHEET

Fill in the blanks to expand each expression.

1 $(a + 4)^2 = (a + 4)(a + 4)$

$= a(\_\_ + \_\_) + 4(\_\_ + \_\_)$

$= a^2 + 4a + 4a + \_\_$

$(a + \underline{4})^2 = a^2 + 8a + \_\_$

2 $(p + 3)^2 = (p + 3)(p + 3)$

$= p(\_\_ + \_\_) + 3(\_\_ + \_\_)$

$= p^2 + \_\_ p + \_\_ p + \_\_$

$(p + \underline{3})^2 = p^2 + \_\_ p + \_\_$

Can you see a relationship between each of these numbers and the underlined number in the brackets?

3 $(m - 5)^2 = (m - 5)(m - 5)$

$= m(\_\_ - \_\_) - 5(\_\_ - \_\_)$

$= m^2 - 5m - 5m + \_\_$

$(m - \underline{5})^2 = m^2 - \_\_ m + 25$

4 $(k - 8)^2 = (k - 8)(k - 8)$

$= k(k - 8) - \_\_ (k - 8)$

$= k^2 - \_\_ k - \_\_ k + \_\_$

$(k - \underline{8})^2 = k^2 - \_\_ k + \_\_$

Can you see a relationship between each of these numbers and the underlined numbers in the brackets?

5 $(3p - 7)^2 = (3p - 7)(3p - 7)$

$= 3p(3p - 7) - \_\_ (3p - 7)$

$= 9p^2 - 21p - 21p + \_\_$

$(\underline{3}p - \underline{7})^2 = 9p^2 - \_\_ p + \_\_$

6 $(5m + 3y)^2 = (5m + 3y)(5m + 3y)$

$= 5m(\_\_ + \_\_) + 3y(\_\_ + \_\_)$

$= 25m^2 + \_\_ my + \_\_ my + \_\_ y^2$

$(\underline{5}m + \underline{3}y)^2 = 25m^2 + \_\_ my + \_\_ y^2$

Can you see a relationship between each of these numbers and the underlined numbers in the brackets?

7 $(f - 4)(f + 4) = (f\_\_ + \_\_) - 4(\_\_ + \_\_)$

$= f^2 + 4f - 4f - \_\_$

$(f - \underline{4})(f + \underline{4}) = f^2 - \_\_$

8 $(g + 3)(g - 3) = g(\_\_ - \_\_) + 3(\_\_ - \_\_)$

$= g^2 - \_\_ g + \_\_ g - \_\_$

$(g + \underline{3})(g - \underline{3}) = g^2 - \_\_$

Can you see a relationship between each of these numbers and the underlined numbers in the brackets?

9780170454551

9 $(3p + 7)(3p - 7) = 3p(\_\_\_ - \_\_\_)$
$+ 7(\_\_\_ - \_\_\_)$

$= 9p^2 - 21p + 21p - \_\_\_$

$(\underline{3}p + \underline{7})(\underline{3}p - \underline{7}) = 9p^2 - \_\_\_$

10 $(9y - 5)(9y + 5) = 9y(\_\_\_ + \_\_\_)$
$- 5(\_\_\_ + \_\_\_)$

$= \_\_\_ y^2 + \_\_\_ y - \_\_\_ y - \_\_\_$

$(\underline{9}y - \underline{5})(\underline{9}y + \underline{5}) = \_\_\_ y^2 - \_\_\_$

Can you see a relationship between each of these numbers and the underlined numbers in the brackets?

Use the relationships you have found in the previous questions to answer these without performing the expansion:

a $(w + 6)^2 = \_\_\_^2 + \_\_\_ w + \_\_\_$

b $(k - 4)^2 =$

c $(d + 5)(d - 5) = \_\_\_^2 - \_\_\_$

d $(6f - 7)(6f + 7) = \_\_\_ f^2 - \_\_\_$

e $(2m + 5y)(2m - 5y) = \_\_\_ m^2 - \_\_\_ y^2$

f $(8t - 3)(8t + 3) =$

g $(3m + 1)^2 = \_\_\_ m^2 + \_\_\_ m + \_\_\_$

h $(5m + 7)^2 =$

i $(10t + 3a)^2 \_\_\_^2 + \_\_\_ at \_\_\_ a^2$

j $(2k - 5x)^2 = \_\_\_ k^2 - \_\_\_ kx + \_\_\_ x^2$

k $(8 - 5m)^2 = \_\_\_ - \_\_\_ m + \_\_\_ m^2$

l $(4 - 3g)^2 =$

m $(3y - 10)^2 =$

n $(4a - 11b)^2 =$

12 FACTORISING PUZZLE

MATCH EACH EXPRESSION WITH ITS FACTORISATION NEXT PAGE TO DECODE THE RIDDLE.

25	24	7	5		10	27		18	24	16		21	10	30	30	16	3	26	21	28	16

11	16	18	25	16	26	21		14		1	22	28	5	22	3		7	21	1	

14	21		7	4	17	26	11	3	14		12	5	13	1	16	21	18	?

18	24	16		1	22	28	5	22	3		3	26	28	5	10	30	10	16	27

13	12		25	24	10	4	26		5	24	16		27	18	13	1	26	1	5

30	7	28	18	22	3	10	27	16	12	.

The numbers in the grid above match the question numbers. Factorise each expression and match it with an answer from the 'Key' provided on the next page. Fill in the grid above with the letters that match the questions, to decode a riddle.

1 $3x^2 + 3x - 6$

2 $2x^2 - x - 10$

3 $x^2 + 5x + 4$

4 $x^2 - 64$

5 $2x^2 + 7x - 4$

6 $3x^2 - 6x + 21$

7 $x^2 - 3x - 28$

8 $5x^2 - 10x - 120$

9 $-2x^2 + 72$

10 $-3x^2 + 12x - 9$

11 $3d^2 - 24d + 48$

12 $d^2 - 7d + 10$

13 $d^2 - 16d + 64$

14 $4d^2 - 4d - 15$

15 $2d^2 - 6d - 20$

16 $2d^2 - 13d - 7$

17 $4d^2 + 6d - 14$

18 $2d^2 - 9d + 10$

19 $4d^2 - 24d + 36$

20 $d^2 - 100$

21 $n^2 + 3n - 54$

22 $5n^2 + 8n + 3$

23 $3n^2 - 8n - 16$

24 $2n^2 - 36n + 162$

25 $n^2 - 11n + 24$

26 $8n^2 - 2$

27 $3n^2 - 7n - 6$

28 $5n^2 + 35n + 60$

29 $4n^2 - 16$

30 $48n^2 - 75$

PS PUZZLE SHEET

Key

A	$(x - 7)(x + 4)$	**I**	$-3(x - 1)(x - 3)$	**S**	$(d - 5)(d - 2)$
A	$(2d - 5)(2d + 3)$	**J**	$4(d - 3)^2$	**S**	$(n + 3)(3n - 2)$
B	$3(d - 4)^2$	**K**	$(2x - 5)(x + 2)$	**T**	$(x + 4)(2x - 1)$
C	$5(n + 3)(n + 4)$	**L**	$(x + 8)(x - 8)$	**T**	$(2d - 5)(d - 2)$
D	$3(x + 2)(x - 1)$	**M**	$2(d + 2)(d - 5)$	**U**	$(d - 8)^2$
E	$(2d + 1)(d - 7)$	**N**	$(n - 6)(n + 9)$	**V**	$4(n + 2)(n - 2)$
E	$2(2n + 1)(2n - 1)$	**O**	$(5n + 3)(n + 1)$	**W**	$(n - 3)(n - 8)$
F	$3(4n + 5)(4n - 5)$	**P**	$(3n + 4)(n - 4)$	**X**	$3(x^2 - 2x + 7)$
G	$2(2d^2 + 3d - 7)$	**Q**	$-2(x + 6)(x - 6)$	**Y**	$(d + 10)(d - 10)$
H	$2(n - 9)^2$	**R**	$(x + 1)(x + 4)$	**Z**	$5(x + 4)(x - 6)$

12 SPECIAL BINOMIAL PRODUCTS

Name:

Due date:

Parent's signature:

Part A	/ 8 marks
Part B	/ 8 marks
Part C	/ 8 marks
Part D	/ 8 marks
Total	/ 32 marks

PART A: MENTAL MATHS

1 Evaluate $(-3)^3$.

2 Test whether {20, 21, 29} is a Pythagorean triad.

3 Evaluate $8 - 3b^2$ if $b = -2$.

4 Convert $\frac{1}{6}$ to a recurring decimal.

5 Write the bearing of B from A.

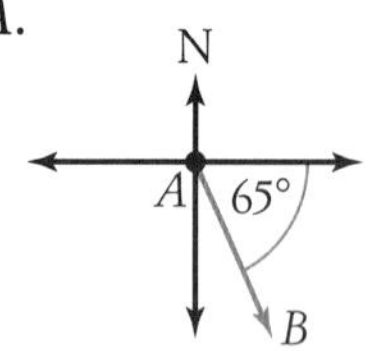

6 Write a simplified algebraic expression for the perimeter of the shape.

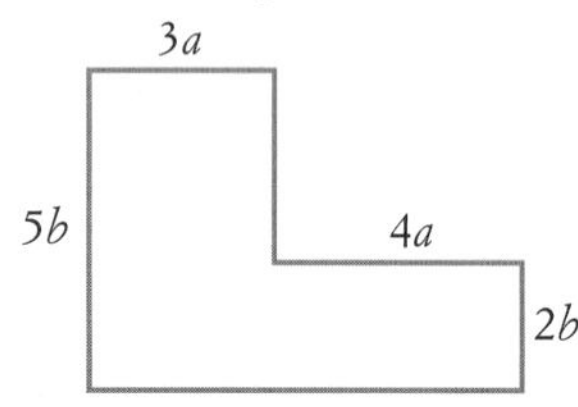

7 Find m and y.

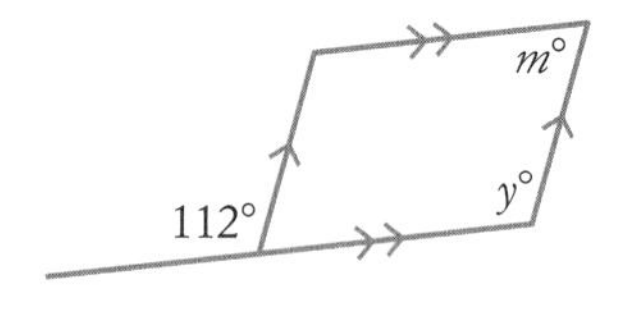

PART B: REVIEW

1 (4 marks) Expand each expression.

a $(p + 4)(p - 1)$

b $(y - 9)(y - 3)$

2 (4 marks) Factorise each expression.

a $x^2 + 8x - 33$

b $a^2 - 11a + 28$

C S F

9780170454551

PART C: PRACTICE

› Special binomial products

1 (4 marks) Expand each expression.

a $(x - 8)^2$

b $(2k + 5)^2$

2 (4 marks) Factorise each expression.

a $2ab + 2da - 3bc - 3dc$

b $6d^2 + d - 2$

c $4y^2 - 81$

PART D: NUMERACY AND LITERACY

1 Write an algebraic expression that is:

a a perfect square

b a difference of 2 squares

2 (4 marks) Expand each expression.

a $(3y - 7)(3y + 7)$

b $(3y - 7)^2$

3 (2 marks) Factorise $3c^2 - 12c - 36$.

HW HOMEWORK

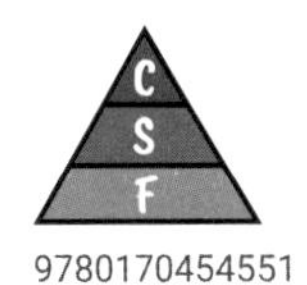

9780170454551

13 STARTUP ASSIGNMENT 13

THIS CHAPTER IS ALL ABOUT SURDS, WHICH ARE SQUARE ROOTS THAT DON'T HAVE AN EXACT VALUE.

PART A: BASIC SKILLS

/ 15 marks

1 Find the median of:

3, 8, 5, 12, 8, 2 ______

2 Expand and simplify:

$2(a + 5) - (3 - a)$

______

______

3

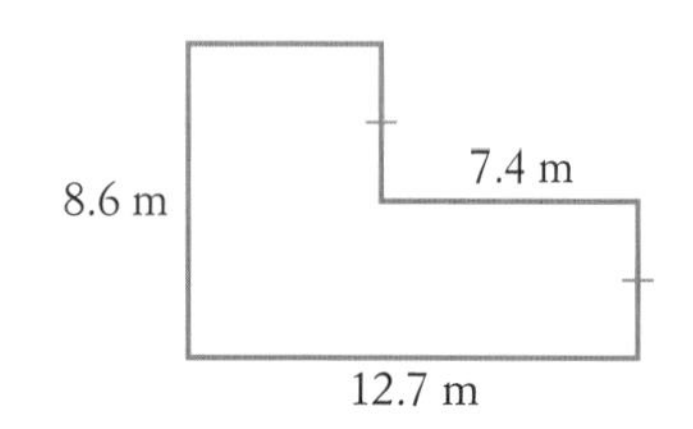

For this figure, calculate:

a the perimeter ______

b the area. ______

______

4 $6.2\text{ m}^2 =$ ______ cm^2

5 85% of what amount is \$531.25? ______

6 Evaluate $\dfrac{2.5 \times 10^5}{\sqrt{15\,625}}$. ______

7 Write an expression for the cost (in dollars) of n tickets if one ticket costs $\$P$.

______

8 Find the value of b in the diagram below.

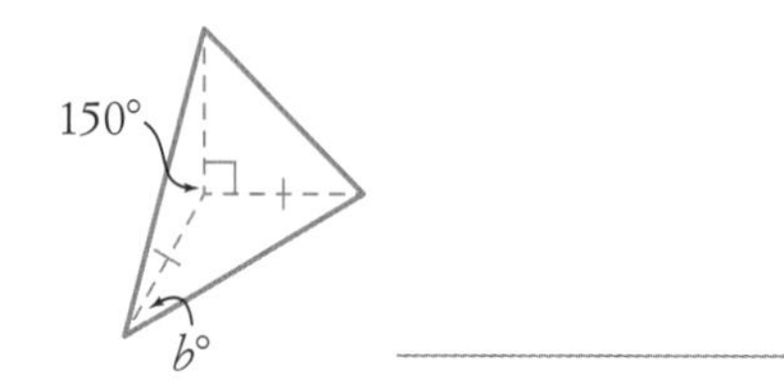

______

9 What is the size of each angle in a regular pentagon? ______

10 Solve $5x - 15 = 3x + 7$. ______

______

______

11 What is the probability that Deni rolls two 6s on a pair of dice? ______

12 Find the midpoint of the interval joining $(1, -4)$ and $(5, 0)$. ______

13 Is $y = -3$ parallel to the x-axis or the y-axis?

______

14 Describe the graph of $x^2 + y^2 = 9$.

______

______

PART B: SURDS AND ALGEBRA

/ 25 marks

15 Complete: $\sqrt{7^2} =$ ______

16 Evaluate $\sqrt{13}$ to 3 significant figures.

______

17 Complete: $\sqrt{2} \times \sqrt{2} =$ ______

18 Circle the rational numbers:

$\sqrt{25}$ $\sqrt{7}$ $\sqrt{15}$ π $0.1\dot{6}$ 3.51

19 Simplify $3x^2 + 2x - x^2 + 3x$. ______

20 Simplify $3a^2 \times 6a^2$. ______

C S F

9780170454551

21 Between which 2 consecutive integers must $\sqrt{76}$ lie? ______

22 Complete: $\sqrt{25 \times 36} =$ ______

23 Expand $3p(r - p)$. ______

24 Find 2 square numbers with a product of:

a 64 ______

b 1225 ______

25 In the diagram below, find h in surd form.

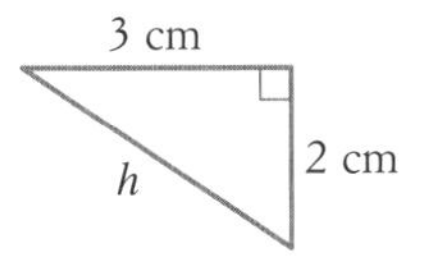

______

26 Expand $(2x - 1)(x + 4)$.

______

______

27 Arrange these from smallest to largest:
$\sqrt{98}$, 3.6^2, $\sqrt{81}$, 10.1, $\sqrt{40}$

______

28 Complete: $(\sqrt{a})^2 =$ ______

29 Find a square number and an integer that have a product of:

a 75 ______

b 80 ______

30 Is it always true that $\sqrt{a + b} = \sqrt{a} + \sqrt{b}$?

______

31 Expand $(2k - 5)(2k + 5)$.

______

32 Is this triangle right-angled?

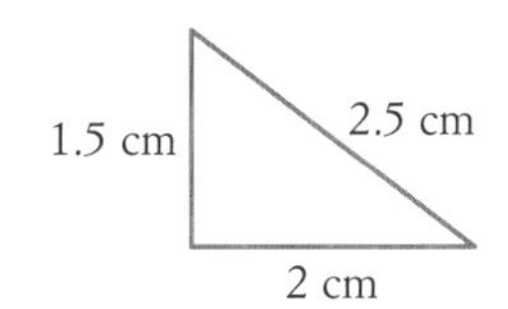

33 **a** Find x in surd form.

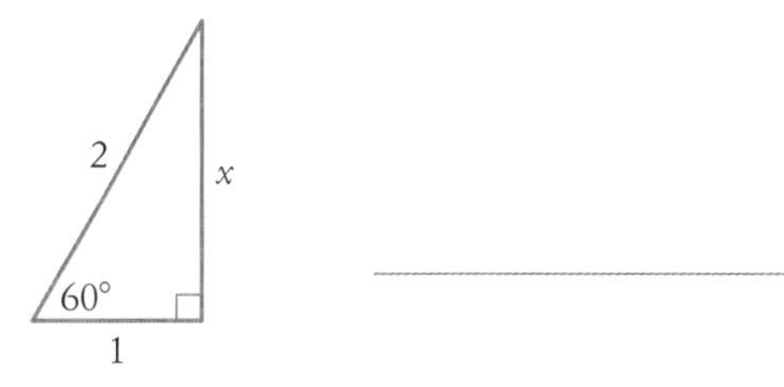

______

b Write sin 60° as a fraction. ______

34 Expand $(3x - y)^2$.

______

35 Is it always true that $\sqrt{a \times b} = \sqrt{a} \times \sqrt{b}$?

______

36 Find the distance between $(-2, 3)$ and $(5, 6)$ on the number plane (in surd form).

______

______

______

PART C: CHALLENGE

Bonus / 3 marks

If a sheet of A4 paper is cut in half, the half-sheet is similar to the whole sheet (that is, matching sides are in the same ratio). If the width of the whole sheet is 1 unit, find x, the length of the sheet, as a surd.

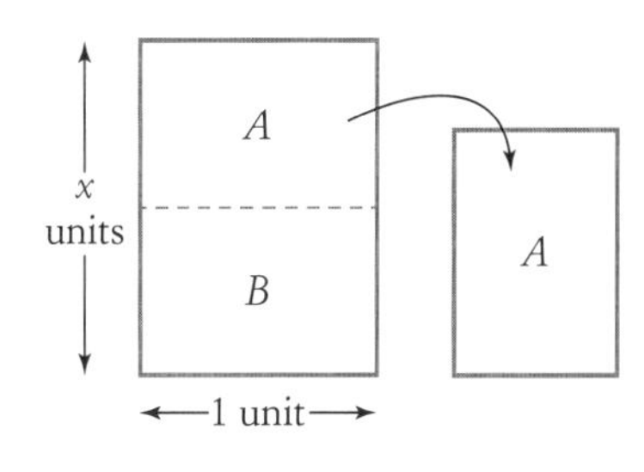

WS WORKSHEET

13 SURDS

HERE'S AN INTRODUCTION TO SURDS.

In the 5th century BCE, the ancient Greek mathematician Pythagoras believed that all things in the world could be explained in terms of **rational numbers** expressed in fraction form $\frac{a}{b}$. He was therefore shocked when he calculated that the length of a diagonal on this square block was $\sqrt{2}$ units, a value that is not rational.

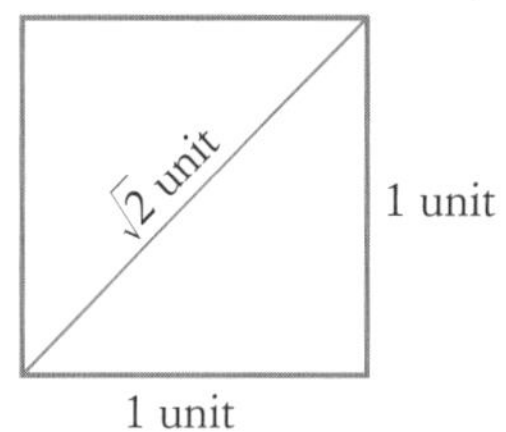

Pythagoras kept this discovery a secret at first, but eventually he called $\sqrt{2}$ an **irrational number** or a **surd** because he believed these new type of numbers were absurd!

There is no exact rational number that if squared is equal to 2. However, $\sqrt{2}$ must lie somewhere between 1 and 2 because $1^2 = 1$ and $2^2 = 4$ and 2 is between 1 and 4. It is impossible to find the exact value of a surd. When written as a decimal, its digits run forever *but without repeating*, for example:

$\sqrt{2} = 1.414\ 213\ 562\ ...$ is **irrational** but $\frac{3}{11} = 0.272727\ ...$ is **rational**.

1 In each list, circle all of the numbers that are irrational.

a $\sqrt{7}, 5, 0, \sqrt{3}, \frac{1}{4}, \frac{11}{20}$

b $\sqrt{4}, \sqrt[3]{100}, \frac{2}{3} \sqrt{10}, 7, 8^3$

c $\pi, \sqrt[3]{125}, 0.3, -4.45, 0.219\ 37...$

d $-1, 0.142\ 857, \sqrt{28}, \sqrt[3]{-27}, 0.04, \sqrt{16}$

2 Complete this list of square numbers:

1, 4, 9, _____, _____, _____, _____, _____, _____, _____

3 Without using a calculator, complete the blanks with consecutive whole numbers.

a $\sqrt{20}$ is between _____ and _____.

b $\sqrt{85}$ is between _____ and _____.

c $\sqrt{42}$ is between _____ and _____.

d $\sqrt{50}$ is between _____ and _____.

e $\sqrt{12}$ is between _____ and _____.

f $\sqrt{33}$ is between _____ and _____.

4 Evaluate each expression.

a $\sqrt{25} \times \sqrt{25}$ _____

b $\sqrt{7} \times \sqrt{7}$ _____

c $\sqrt{10} \times \sqrt{10}$ _____

d $\sqrt{8} \times \sqrt{8}$ _____

e $\sqrt{12} \times \sqrt{12}$ _____

f $(\sqrt{4})^2$ _____

g $(\sqrt{11})^2$ _____

h $(\sqrt{2})^2$ _____

$(\sqrt{a})^2 = \sqrt{a} \times \sqrt{a} = a$

5 Evaluate each expression.

a $\sqrt{4} \times \sqrt{25}$ _____

b $\sqrt{4 \times 25}$ _____

c $\sqrt{100} \times \sqrt{9}$ _____

d $\sqrt{100 \times 9}$ _____

e $\sqrt{36} \times \sqrt{9}$ _____

9780170454551

f $\sqrt{36 \times 9}$ ______

g $\sqrt{16} \times \sqrt{49}$ ______

h $\sqrt{16 \times 49}$ ______

i $\sqrt{3} \times \sqrt{27}$ ______

j $\sqrt{3 \times 27}$ ______

$$\sqrt{a} \times \sqrt{b} = \sqrt{ab}$$

SIMPLIFYING SURDS

Even if we don't know the exact value of a surd, we can use the rule $\sqrt{ab} = \sqrt{a} \times \sqrt{b}$ to simplify a surd if one of its factors is a square number. Study each example.

a $\sqrt{50} = \sqrt{25} \times \sqrt{2} = 5\sqrt{2}$

b $\sqrt{32} = \sqrt{16} \times \sqrt{2} = 4\sqrt{2}$

c $\sqrt{27} = \sqrt{9} \times \sqrt{3} = 3\sqrt{3}$

d $\sqrt{20} = \sqrt{4} \times \sqrt{5} = 2\sqrt{5}$

Make sure you use the highest square factor.

6 Simplify each surd. Check that your answers are correct using a calculator.

a $\sqrt{45}$ ______

b $\sqrt{18}$ ______

c $\sqrt{48}$ ______

d $\sqrt{300}$ ______

e $\sqrt{75}$ ______

f $\sqrt{12}$ ______

g $\sqrt{8}$ ______

h $\sqrt{98}$ ______

i $\sqrt{24}$ ______

j $\sqrt{80}$ ______

k $\sqrt{108}$ ______

l $\sqrt{28}$ ______

ADDING AND SUBTRACTING SURDS

Like algebraic terms, 'like surds' can be added and subtracted. Study each example.

a $3\sqrt{2} + 4\sqrt{2} = 7\sqrt{2}$

b $\sqrt{150} - 3\sqrt{6} = \sqrt{25} \times \sqrt{6} - 3\sqrt{6}$

$= 5\sqrt{6} - 3\sqrt{6}$

$= 2\sqrt{6}$

7 Simplify each expression. Check that your answers are correct using a calculator.

a $10\sqrt{7} + \sqrt{7}$ ______

b $8\sqrt{3} - 2\sqrt{3}$ ______

c $6\sqrt{5} + 2\sqrt{5} - 4\sqrt{5}$ ______

d $9\sqrt{11} - 8\sqrt{11}$ ______

e $2\sqrt{6} + 2\sqrt{10} - 3\sqrt{10} + 7\sqrt{6}$ ______

f $4\sqrt{2} + \sqrt{3} + 2\sqrt{2} - 3\sqrt{3}$ ______

g $\sqrt{8} + 3\sqrt{2}$ ______

h $2\sqrt{3} + 3\sqrt{12}$ ______

i $\sqrt{45} - \sqrt{5}$ ______

j $\sqrt{75} - \sqrt{27}$ ______

C S F

13 SIMPLIFYING SURDS

First complete this pattern of the first 10 square numbers:

1, 4, 9, ______, ______, ______, ______, ______, ______, ______

Simplify the surds in the puzzle below and draw straight lines between the dots to join equivalent surds. Each correct line you draw will go through one letter and one number and these pairs are the solution to the code. Write the numbers corresponding to the letters given in the grid under the puzzle to reveal a famous irrational number.

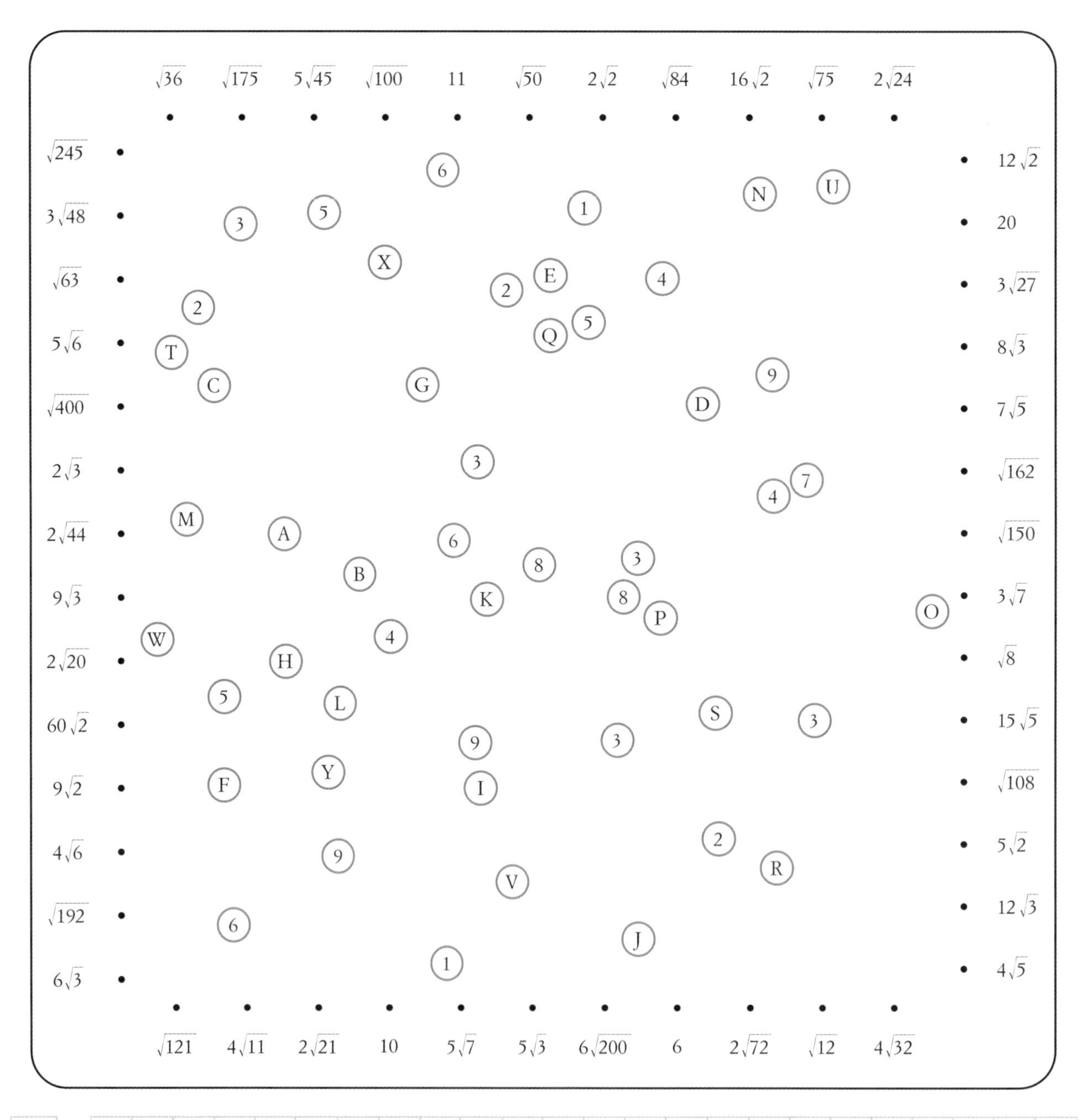

A		B	C	D	E	F	G	H	I	J	K	L	M	N	O	P	Q	R	S	T	U	V	W	X	Y
	.																								

RATIONALISING THE DENOMINATOR 13

WORKSHEET

Rationalise the denominator of each surdic expression.

1 $\dfrac{5}{\sqrt{2}}$ ____________

2 $\dfrac{3}{\sqrt{6}}$ ____________

3 $\dfrac{\sqrt{5}}{\sqrt{3}}$ ____________

4 $\dfrac{\sqrt{10}}{2\sqrt{5}}$ ____________

5 $\dfrac{2\sqrt{3}}{\sqrt{2}}$ ____________

6 $\dfrac{5\sqrt{5}}{2\sqrt{6}}$ ____________

7 $\dfrac{2+\sqrt{3}}{\sqrt{5}}$ ____________

8 $\dfrac{\sqrt{5}-3}{\sqrt{6}}$ ____________

9 $\dfrac{6+\sqrt{12}}{\sqrt{3}}$ ____________

10 $\dfrac{7-\sqrt{5}}{3\sqrt{5}}$ ____________

11 $\dfrac{4\sqrt{5}-2\sqrt{3}}{\sqrt{2}}$ ____________

12 $\dfrac{7\sqrt{10}+3\sqrt{7}}{\sqrt{7}}$ ____________

13 $\dfrac{8-\sqrt{6}}{2\sqrt{2}}$ ____________

14 $\dfrac{7+3\sqrt{5}}{3\sqrt{10}}$ ____________

Extension

Hint: For question **15**, multiply the numerator and denominator by $(5-\sqrt{2})$.

15 $\dfrac{6}{5+\sqrt{2}}$ ____________

16 $\dfrac{4}{\sqrt{5}-1}$ ____________

17 $\dfrac{\sqrt{3}}{\sqrt{6}+2}$ ____________

18 $\dfrac{4\sqrt{6}}{5-\sqrt{5}}$ ____________

19 $\dfrac{3\sqrt{2}}{2\sqrt{3}-3}$ ____________

20 $\dfrac{6+\sqrt{2}}{4-\sqrt{2}}$ ____________

Mixed answers: $\sqrt{70}+3$, $\dfrac{2\sqrt{5}+\sqrt{15}}{5}$, $\dfrac{5\sqrt{30}}{12}$, $\dfrac{5\sqrt{2}}{2}$, $\sqrt{5}+1$, $\dfrac{3\sqrt{2}-2\sqrt{3}}{2}$, $\dfrac{7\sqrt{5}-5}{15}$, $\dfrac{\sqrt{15}}{3}$, $\sqrt{6}$, $2\sqrt{6}+3\sqrt{2}$, $\dfrac{4\sqrt{2}-\sqrt{3}}{2}$, $\dfrac{5\sqrt{6}+\sqrt{30}}{5}$, $\dfrac{7\sqrt{10}+15\sqrt{2}}{30}$, $2\sqrt{3}+2$, $\dfrac{10-5\sqrt{2}}{3}$, $\dfrac{\sqrt{30}-3\sqrt{6}}{6}$, $\dfrac{\sqrt{2}}{2}$, $\dfrac{13+5\sqrt{2}}{7}$, $2\sqrt{10}-\sqrt{6}$, $\dfrac{30-6\sqrt{2}}{23}$, $\dfrac{\sqrt{6}}{2}$

13 SURDS 1

A SURD IS IRRATIONAL, AND ABSURD MEANS 'MAKES NO SENSE'. CAN YOU SEE HOW THESE 2 WORDS ARE RELATED?

Name:

Due date:

Parent's signature:

Part	Marks
Part A	/ 8 marks
Part B	/ 8 marks
Part C	/ 8 marks
Part D	/ 8 marks
Total	/ 32 marks

PART A: MENTAL MATHS

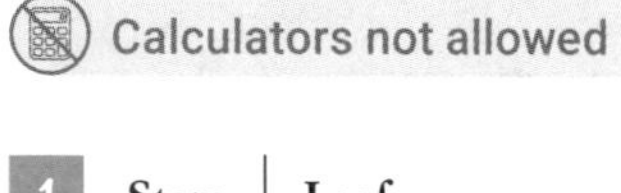

Calculators not allowed

1

Stem	Leaf
1	3 6
2	1 4 4 7
3	0 2 2 3
4	4 7 8

For the above data, find:

a the mode ____________

b the median ____________

2 Solve $2m + 8 = 3(2m - 5)$.

____

____

3 Find x.

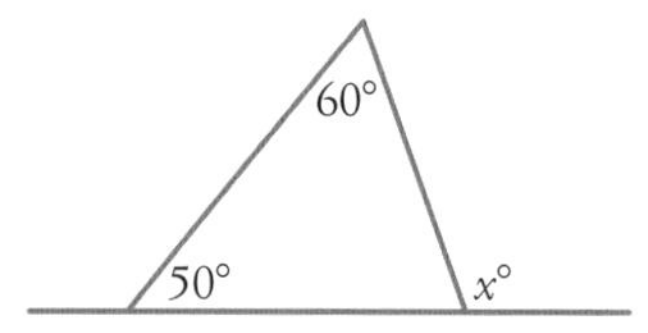

____

4 What is the centre and radius of the circle with equation $x^2 + y^2 = 36$?

____

5 Find the surface area of this prism.

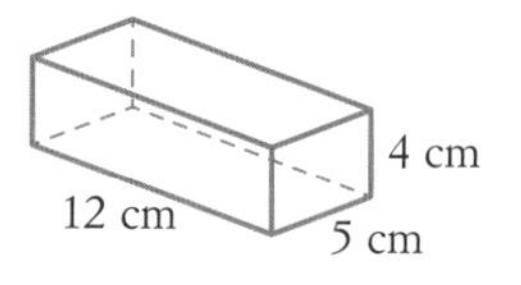

____

____

____

6 Find the midpoint of the interval joining points (4, 8) and (−2, 6) on the number plane.

____

7 Is the age you turn this year discrete or continuous data? ____________

PART B: REVIEW

1 Circle the square numbers in this list:

22 49 75 32 100

____

2 Expand each expression.

a $-6(3p + 4)$

____

9780170454551

b $(x-3)(x-7)$

3 Simplify each expression.

a $\left(\sqrt{x}\right)^2$

b $\sqrt{5^2}$

4 Circle the surds in this list:

$\sqrt{49}$ $\sqrt{88}$ $\sqrt{81}$ $\sqrt{125}$ $\sqrt{144}$

5 Is each number rational or irrational?

a $\sqrt{99}$

b $\sqrt{289}$

PART C: PRACTICE

› Symplifying surds
› Adding and subtracting surds

1 Simplify each expression.

a $\sqrt{160}$

b $\dfrac{\sqrt{250}}{50}$

c $9\sqrt{3}-4\sqrt{3}$

d $4\sqrt{6}-5\sqrt{5}+6\sqrt{6}$

2 (4 marks) Simplify each expression.

a $\sqrt{27}-\sqrt{8}+\sqrt{18}$

b $5\sqrt{20}-3\sqrt{125}$

PART D: NUMERACY AND LITERACY

1 True or false?

a $\sqrt{x}$ has 2 answers when x is positive

b $\sqrt{x}$ has no value when x is negative

c $\sqrt{x}=x$ for only 2 values of x

d $\sqrt{x}$ is a surd when x is positive

2 Is each number rational (R) or irrational (I)?

a 10π

b $0.7\dot{4}$

c $\sqrt[3]{-8}$

d $2+\sqrt{5}$

HW HOMEWORK

C S F

13 SURDS 2

Name:

Due date:

Parent's signature:

Part A	/ 8 marks
Part B	/ 8 marks
Part C	/ 8 marks
Part D	/ 8 marks
Total	/ 32 marks

I SEE SOME RED QUESTIONS ON THESE PAGES. THIS ALGEBRA NEEDS YOUR CAREFUL BRAIN TRAINING.

PART A: MENTAL MATHS

Calculators not allowed

1 (2 marks) Find the area of this triangle.

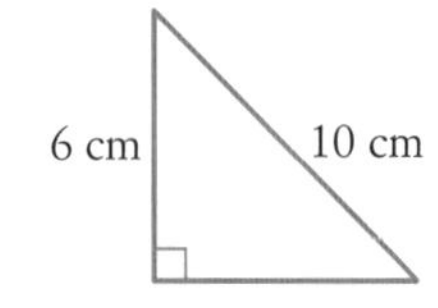

2 Find the gradient of this line.

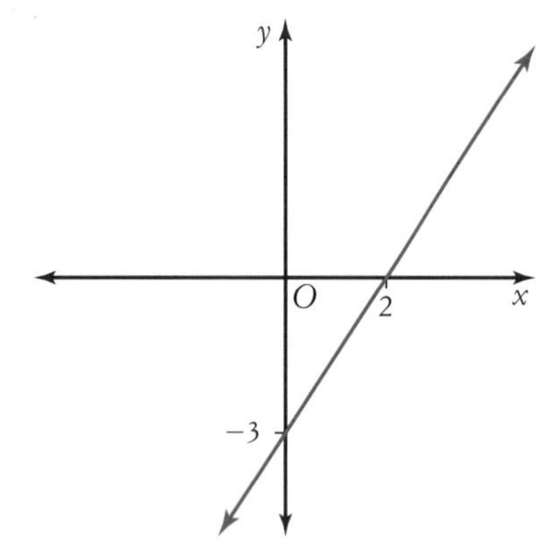

3 Simplify 36 : 48 : 18.

4 Expand and simplify $3(1 - 2x) - x(2 - x)$.

5 Factorise $49x^2 - 42x + 9$.

6 (2 marks) 2 coins are thrown together. List all the possible pairings in the sample space (H = heads, T = tails).

PART B: REVIEW

1 Simplify each surd.

a $\sqrt{275}$

b $7\sqrt{48}$

2 (6 marks) Simplify each expression.

a $\sqrt{200} + \sqrt{18}$

C S F

HOMEWORK HW

9780170454551

b $3\sqrt{5}+\sqrt{50}-2\sqrt{125}$

c $\sqrt{45}-3\sqrt{63}+5\sqrt{80}$

PART C: PRACTICE

› Multiplying and dividing surds
› Rationalising the denominator

1 Simplify each expression.

a $5\sqrt{10}\times 3\sqrt{3}$

b $\dfrac{\sqrt{128}}{\sqrt{2}}$

c $\sqrt{80}\div\left(-4\sqrt{5}\right)$

2 (5 marks) Rationalise the denominator of each surd.

a $\dfrac{5}{\sqrt{2}}$

b $\dfrac{14}{3\sqrt{7}}$

c $\dfrac{5-\sqrt{3}}{2\sqrt{3}}$

PART D: NUMERACY AND LITERACY

1 a What is the first step of rationalising the denominator of $\dfrac{20}{\sqrt{5}}$?

b Why is it called 'rationalising the denominator'?

2 Complete each formula.

a $\sqrt{xy}=$

b $\sqrt{\dfrac{x}{y}}=$

c $(a-b)^2=$

3 (3 marks) Expand and simplify each expression.

a $\left(\sqrt{5}-2\right)^2$

b $\left(5\sqrt{2}-4\right)\left(\sqrt{2}+5\right)$

HW HOMEWORK

C S F

13 SURDS CROSSWORD

THOSE ARE THE WORDS, HERE IS THE PUZZLE. YOU KNOW WHAT TO DO.

The answers to this crossword are listed below, in alphabetical order.

Arrange them in the correct places in the puzzle.

ANNULUS	AREA	BASE	CAPACITY
APPROXIMATE	BINOMIAL	DENOMINATOR	DIFFERENCE
EXPAND	EXPRESSION	IRRATIONAL	PRODUCT
QUOTIENT	RATIONAL	RATIONALISE	REAL
ROOT	SIMPLIFY	SQUARE	SURD
UNDEFINED			

PS PUZZLE SHEET

STARTUP ASSIGNMENT 14 (14)

WE'LL BE COVERING QUADRATIC EQUATIONS AND THE PARABOLA IN THIS TOPIC.

WS WORKSHEET

PART A: BASIC SKILLS / 15 marks

1 Calculate $\cos 45^\circ$, correct to 3 significant figures. ______

2 Evaluate $4^{\frac{1}{2}} \times 4^{-2}$. ______

3 Complete: $10\text{ m}^3 =$ ______ cm^3.

4 Complete: $55\text{ m/s} =$ ______ km/h.

5 Write a possible equation for this line.

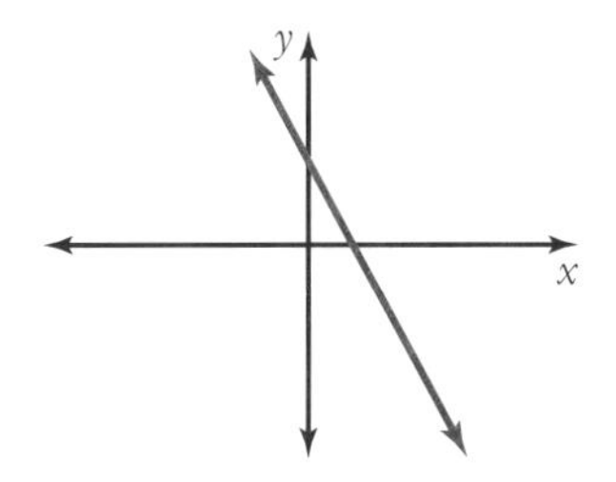

______

6 Simplify $\dfrac{20x^3y^2}{4x^3y}$. ______

7 Find the surface area of a rectangular prism that has dimensions 4.6 cm, 2.1 cm and 5.0 cm.

______

______

8 Evaluate $\dfrac{2.46 \times 10^{12}}{4 \times 10^8}$. ______

9 Name the quadrilateral that has equal diagonals that bisect each other.

______

10 If $\dfrac{5}{6}$ of a number is 90, what is the number?

______

______

______

11 A number from 1 to 20 is chosen at random. What is the probability that it is a multiple of 3?

______

12 Find θ to the nearest degree if $\sin \theta = 0.561$.

______

13 Divide \$4550 in the ratio 2 : 1 : 4.

______

14 What is the sum of the exterior angles of any polygon?

______

15 Madeline is paid \$465 plus a 5.5% commission on any jewellery she sells. Calculate her pay if she sells \$3812 worth of jewellery.

______

PART B: ALGEBRA AND GRAPHS / 25 marks

16 For the line $y = 3x - 2$, find:

a its gradient ______

b its y-intercept. ______

17 If $x = 3$, find y if:

a $y = x^2 + 5$ ______

b $y = 2x^2 - x + 7$ ______

c $y = (x - 4)(x + 2)$ ______

d $y = 3x - x^2$ ______

WORKSHEET

18 Graph $y = 3x^2$ on a number plane.

19 If $h = 3t^2 + 7t + 10$, find h if:

a $t = 2$ ____________

b $t = -2$ ____________

c $t = 5$ ____________

20 Solve:

a $x^2 - 5x - 14 = 0$ ____________

____

b $y^2 - 4y - 5 = 0$ ____________

____

21 Expand $(x - 3)^2$.

____

22 Factorise $3x^2 + 11x - 20$.

____

____

23 If $x = -2$, find y if:

a $y = x^2 - 4$ ____________

b $y = -x^2 + 4x - 3$ ____________

c $y = 4x^2 - 6x$ ____________

24 Graph $y = -\frac{1}{2}x^2$ on a number plane.

25

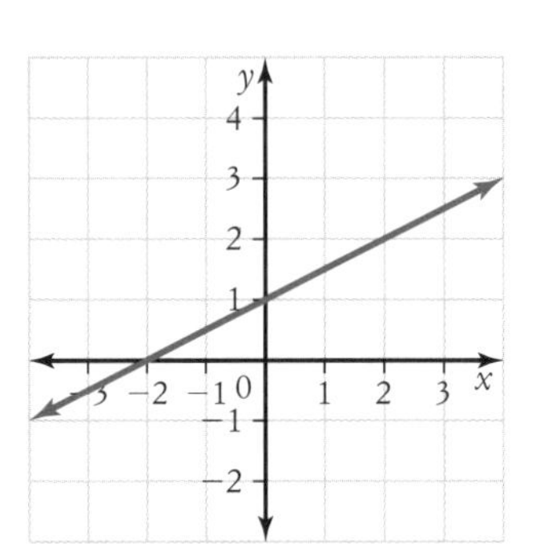

For this line, find:

a its y-intercept ____________

b its x-intercept ____________

c its gradient ____________

d its equation. ____________

26 If $a = -5$ and $b = -2$, evaluate $-\frac{b}{2a}$.

____

27 Evaluate $\sqrt{b^2 - 4ac}$, correct to 2 decimal places, if:

a $a = 2, b = 4$ and $c = 1$ ____________

b $a = -1, b = -3$ and $c = 4$ ____________

PART C: CHALLENGE

Bonus / 3 marks

What is the value of $\sqrt{2 + \sqrt{2 + \sqrt{2 + \sqrt{2 + \ldots}}}}$?

There is a definite answer. Can you prove it?

Hint: Let $x = \sqrt{2 + \sqrt{2 + \sqrt{2 + \sqrt{2 + \ldots}}}}$ and square both sides. ____________

C S F

9780170454551

GRAPHING PARABOLAS 2 (14)

Teacher's tickbox

Graph the ticked set of quadratic equations.

❏ $y = x^2, y = 2x^2, y = \frac{1}{2}x^2, y = -2x^2$

❏ $y = x^2 + 4, y = 2x^2 - 2$

❏ $y = x^2 - 2x - 3, y = -x^2 + 5x - 6$

❏ $y = \frac{1}{2}x^2 - 4x + 1, y = -x^2 - 4x + 5$

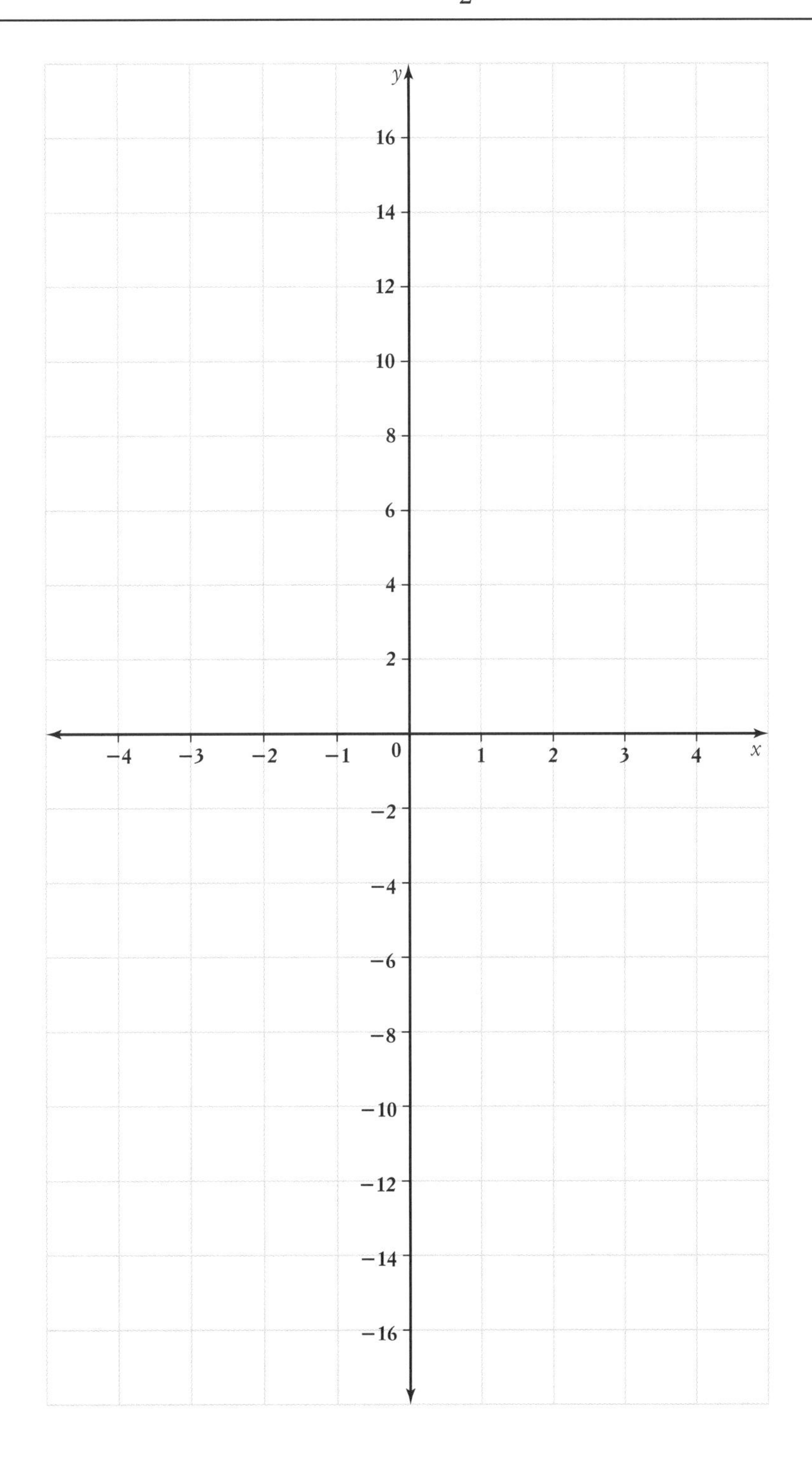

(14) QUADRATIC EQUATIONS PUZZLE

THE SOLUTIONS TO THESE EQUATIONS CAN BE FOUND NEXT PAGE. MATCH THEM TO SOLVE THE RIDDLE.

PS PUZZLE SHEET

12	29	23	26	25	30	2		24	9	18	19		2	23	11	22		18	19	

29	25	27	7		2	28	13	12	9	18	30	2		29	25	30	7	19		23	30	10

6	15	28	8	7	19		21	30		13		30	15	11	20	7	28		12	29	23	30	7 .

26	5	15 ,		8	22		2	21	24		16	5		29	21	5	27		14	5	28

16	9	22		25	30	24	22	28	6	7	12	16	19 .

The numbers in the grid above match the question numbers. Solve each quadratic equation and match the solution with an answer from the 'Key' provided on the next page. Fill in the grid above with the letters that match the questions, to discover the answer to the riddle:

What did the maths coach say to his team of football players?

1 $(x - 3)(x - 2) = 0$

____

2 $(2x - 3)(3x + 1) = 0$

____

3 $x^2 - 16 = 0$

____

4 $2x^2 - 18 = 0$

____

5 $x^2 - 2 = 0$

____

6 $x(2x + 5) = 0$

____

7 $x^2 = 3x$

____

8 $x^2 - 6x + 5 = 0$

____

9 $2x^2 - 10x = 0$

____

10 $3x^2 - 5x - 2 = 0$

____

11 $6x^2 - 5x - 6 = 0$

____

12 $4x^2 - 13x - 12 = 0$

____

13 $x^2 + 7x - 10 = 0$

____

14 $3x^2 - 11x + 6 = 0$

____

15 $x(x + 2) + 3(x + 2) = 0$

____

9780170454551

16 $\frac{x}{3} = \frac{12}{x}$

____

17 $(x + 2)^2 = 5$

____

18 $4x^2 + 2x - 6 = 0$

____

19 $x^2 - 8x + 2 = 0$

____

20 $\frac{16}{x - 3} = x + 3$

____

21 $2x^2 + x - 5 = 0$

____

22 $3x(x - 1) = 2(5 - 2x)$

____

23 $(2x - 1)(3x + 1) = 4$

____

24 $2x^2 + 12x + 10 = 0$

____

25 $(3x - 1)^2 = 100$

____

26 $\frac{x - 3}{7} = \frac{4}{x}$

____

27 $6x - 8x^2 = 1$

____

28 $3x^2 - 6x + 2 = 0$

____

29 $x(2x - 3) = -1$

____

30 $x = \frac{5x + 14}{x}$

____

PUZZLE SHEET

Key

A	$x = -\frac{5}{6}, 1$
A	$x = \frac{-7 \pm \sqrt{89}}{2}$
B	$x = \pm 5$
C	$x = 0, -\frac{5}{2}$
D	$x = -\frac{1}{3}, 2$
E	$x = -2, \frac{5}{3}$
E	$x = 0, 3$
F	$x = \frac{2}{3}, 3$

G	$x = \frac{3}{2}, -\frac{1}{3}$
H	$x = 0, 5$
I	$x = 1, -\frac{3}{2}$
I	$x = \frac{11}{3}, -3$
J	$x = -2 \pm \sqrt{5}$
K	$x = \frac{1}{4}, \frac{1}{2}$
L	$x = \frac{1}{2}, 1$
M	$x = -\frac{2}{3}, \frac{3}{2}$

N	$x = 7, -2$
O	$x = \pm\sqrt{2}$
O	$x = \frac{-1 \pm \sqrt{41}}{4}$
P	$x = 4, -\frac{3}{4}$
Q	$x = \pm 4$
R	$x = \frac{3 \pm \sqrt{3}}{3}$
S	$x = 4 \pm \sqrt{14}$
T	$x = -5, -1$

T	$x = \pm 6$
U	$x = -3, -2$
V	$x = 1, 5$
W	$x = 2, 3$
X	$x = \pm 3$
Y	$x = -4, 7$

QUADRATIC EQUATIONS CROSSWORD

Unscramble the word clues and write them in the crossword.

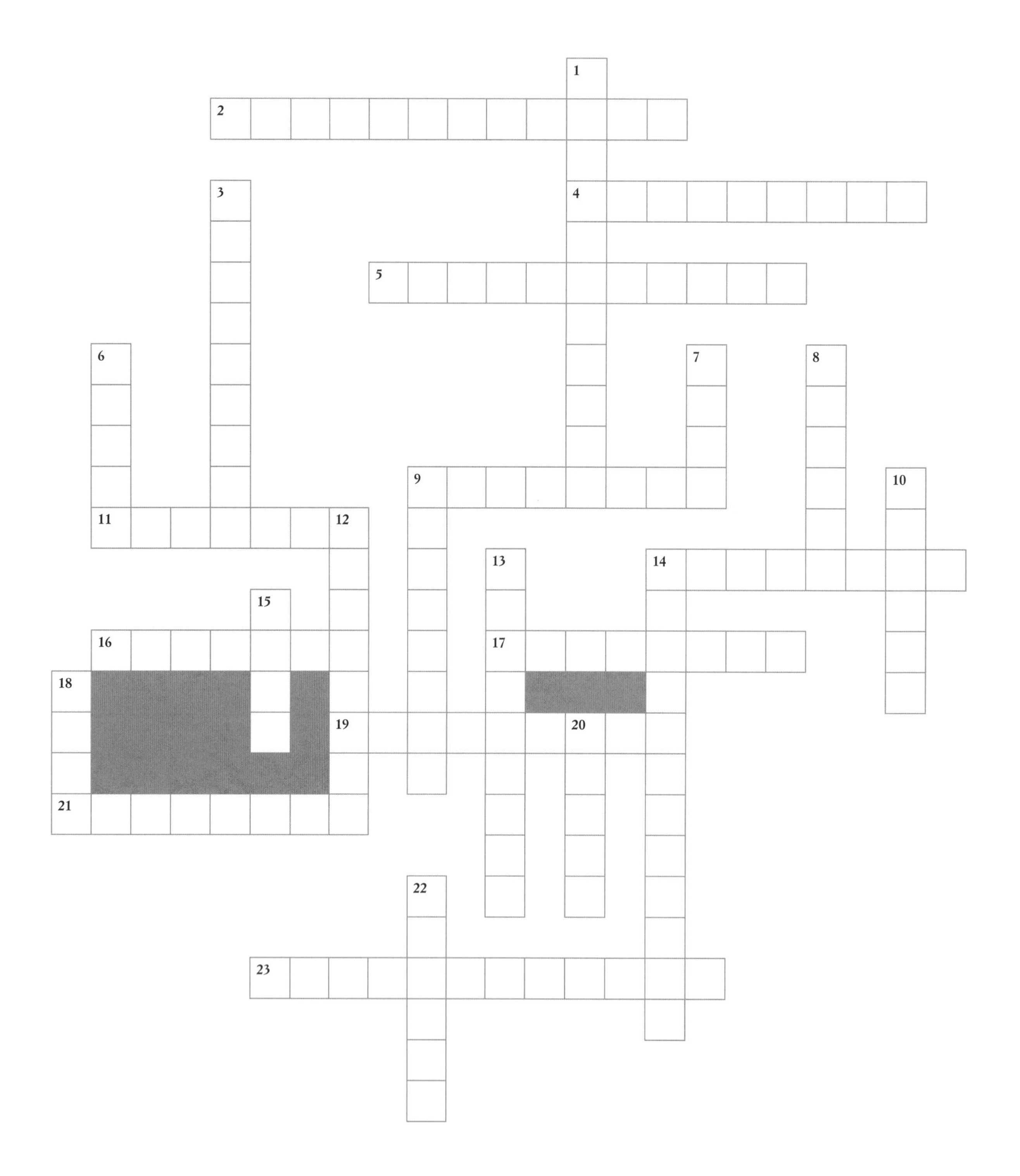

Across

2 TSUNAMILOUSE

4 SEORIFATC

5 TESOCONRDIA

9 STNNACOT

11 CACOEVN

14 MRTMYSYE

16 FLUROAM

17 BALARAPO

19 TIPCENTRE

21 STULOION

23 REECTNIOINST

Down

1 CIEFFOCEITN

3 AADIQCURT

6 CONIM

7 TORO

8 SERQUA

9 LETMECOP

10 XRVETE

12 ANIQUOTE

13 BYHPAOLER

14 NBSTIUSUTTOI

15 RUSD

18 SXAI

20 TAXCE

22 RCCLEI

14 QUADRATIC EQUATIONS

Name:

Due date:

Parent's signature:

Part A	/ 8 marks
Part B	/ 8 marks
Part C	/ 8 marks
Part D	/ 8 marks
Total	/ 32 marks

PART A: MENTAL MATHS

Calculators not allowed

1 An interval AB joins $A(-8, 5)$ and $B(4, 0)$ on the number plane. Find:

a its length

b its gradient

2 (4 marks) Factorise each expression.

a $x^2 + 3x - 88$

b $5n^3 - 125n$

3 For this set of data, find:

6 9 9 11 13 14 15 16 19

a the median

b the interquartile range

PART B: REVIEW

1 Solve each equation.

a $(x + 4)(x - 6) = 0$

b $x(2x - 9) = 0$

2 (6 marks) Solve each equation.

a $(2u - 7)(3u + 5) = 0$

b $4k^2 - 12k = 0$

C S F

HW HOMEWORK

9780170454551

c $7p^2 + 11p - 6 = 0$

PART C: PRACTICE

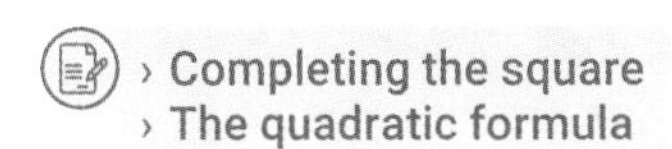

1 Solve each equation, writing the solution in surd form.

a $(x + 3)^2 = 5$

b $\left(y - \frac{1}{2}\right)^2 = \frac{3}{4}$

2 (4 marks) Solve each equation by completing the square.

a $a^2 + 2a - 5 = 0$

b $2b^2 + 4b - 6 = 0$

3 (2 marks) Solve $3x^2 + x - 7 = 0$ using the quadratic formula.

PART D: NUMERACY AND LITERACY

1 Find the numbers that 'complete the square' in this equation.

$$x^2 - 10x + \_\_\_\_ = (x - \_\_\_\_)^2$$

2 (6 marks) Solve each equation.

a $(2x - 5)^2 = 10$

b $5x^2 + 3x - 2 = 0$

c $3x^2 = 8x + 3$

HW HOMEWORK

C S F

14 THE PARABOLA

FOR QUESTION 4, THE SURFACE AREA OF A SPHERE IS $4\pi R^2$, BUT THE SURFACE AREA OF A HEMISPHERE IS NOT $2\pi R^2$.

Name:

Due date:

Parent's signature:

Part A	/ 8 marks
Part B	/ 8 marks
Part C	/ 8 marks
Part D	/ 8 marks
Total	/ 32 marks

PART A: MENTAL MATHS

Calculators not allowed

1 (2 marks) Factorise $6a^2 + 11a + 3$.

2 Make x the subject of the formula $2y = \sqrt{\frac{x}{5}}$.

3 Solve $-2m + 10 < -6$.

4 The formula for the surface area of a sphere is $SA = 4\pi r^2$. Find the surface area of this hemisphere in terms of π.

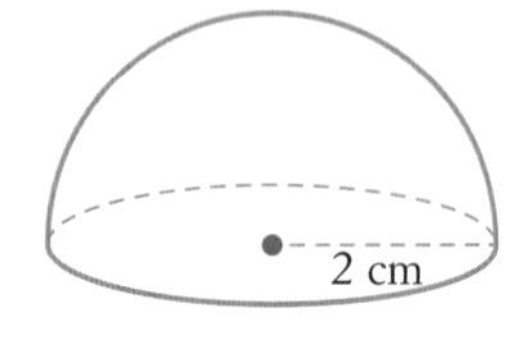

5 (2 marks) Solve these simultaneous equations.

$-3x + 4y = 20$

$6x - 5y = -19$

6 Simplify $\sqrt{147} - \sqrt{75}$.

PART B: REVIEW

1 (3 marks) This right-angled triangle has an area of 600 m². Its base is 10 m longer than its height. Find its base and height.

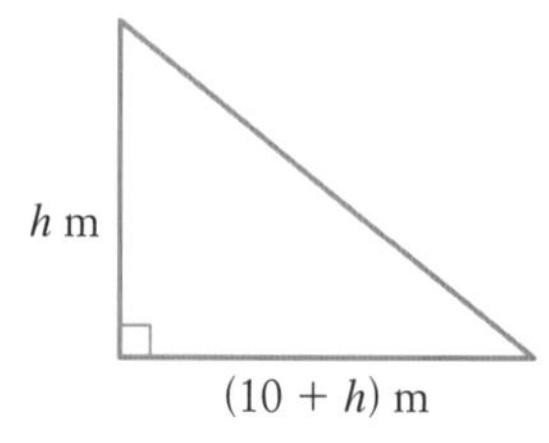

C S F

9780170454551

HW HOMEWORK

2 (2 marks) Solve $3x^2 - 3x - 6 = 0$.

3 (3 marks) The product of 2 consecutive even numbers is 1848. Find the numbers.

PART C: PRACTICE

› The parabola $y = ax^2 + bx + c$
› The axis of symmetry and the vertex of a parabola

1 For this parabola, find:

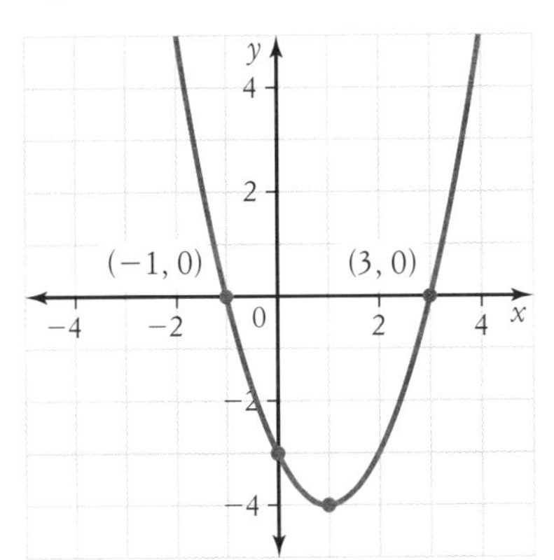

a the equation of the axis of symmetry

b the coordinates of the vertex

2 (6 marks) Graph $y = -2x^2 + 6x - 4$, showing its vertex, x- and y- intercepts.

PART D: NUMERACY AND LITERACY

1 (2 marks) Is the graph of $y = x^2 - 6x + 8$ concave up or concave down? Give reasons.

2 Find the y-intercept of the graph of $y = x^2 - 6x + 8$.

3 Find the equation of the axis of symmetry of the graph of $y = x^2 - 6x + 8$.

4 Find the coordinates of the vertex of the graph of $y = x^2 - 6x + 8$.

5 (2 marks) Find the x-intercepts of the graph of $y = x^2 - 6x + 8$.

6 Graph $y = x^2 - 6x + 8$.

HW HOMEWORK

C S F

15 STARTUP ASSIGNMENT 15

WS WORKSHEET

PART A: BASIC SKILLS

/ 15 marks

1 35% of an amount is \$87.50. What is the amount? ______

2 Evaluate $16^{\frac{3}{4}}$. ______

3 How many hours and minutes will it take a car travelling at 72 km/h to cover 120 km?

______

______

4 Convert $0.\dot{7}\dot{2}$ to a simple fraction.

______

______

______

5 Find the perimeter of this trapezium.

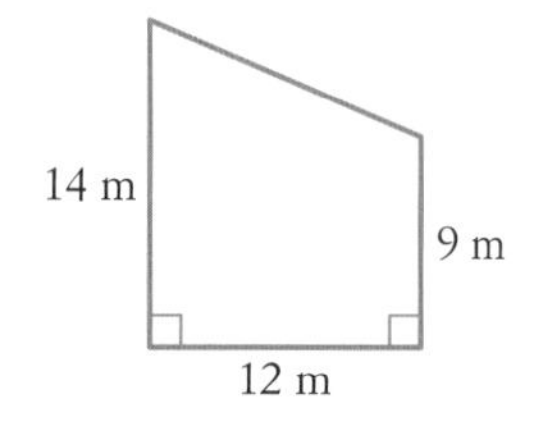

______

______

______

6 Expand and simplify $(3+\sqrt{5})^2$.

______

______

7 If a cube has volume 195.112 cm^3, find its surface area. ______

______

8 Calculate to the nearest cent the compound interest from \$16 000 invested at 6.2% p.a. for 8 years.

______

______

9 Find the gradient of the line perpendicular to $4x - 3y + 1 = 0$.

______

______

10 Factorise $4y^2 - 17y - 15$.

______

______

11 Find d.

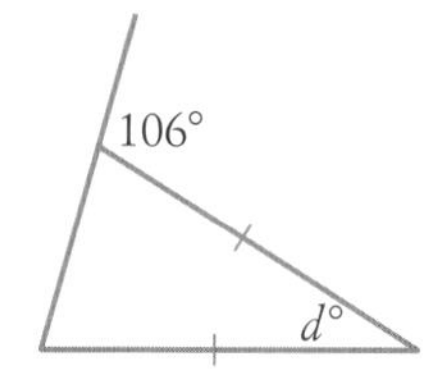

12 Solve $5 + 2k \leq 6 - k$. ______

______

______

13 Find the interquartile range of these numbers: 16, 17, 24, 10, 16, 14, 13, 14, 19

______

14 Solve $-5y^3 = 135$.

______

______

C S F

9780170454551

15 Find the equation of the line passing through (0, 0) and (3, −1).

____

____

PART B: TRIGONOMETRY / 25 marks

16 Write $\cos\theta$ as a fraction. ______________

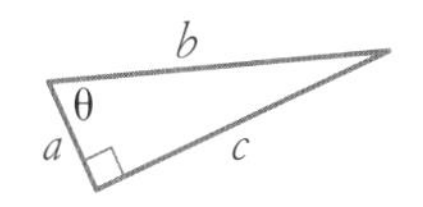

17 Calculate to 4 decimal places:

a $\dfrac{12\sin 70^\circ}{10}$ ______________

b $\dfrac{14\sin 16^\circ}{\sin 42^\circ}$ ______________

c $\dfrac{3^2+5^2-2^2}{2\times3\times5}$ ______________

d $\cos 70^\circ$ ______________

e $\sin 20^\circ$ ______________

18 **a** What do you notice about $\cos 70^\circ$ and $\sin 20^\circ$ in the question above?

____

b $\cos$ ________ $= \sin 50^\circ$

19 Using the triangle, write as a fraction:

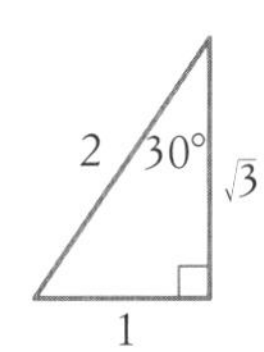

d $\cos 30^\circ$ ______________

e $\tan 30^\circ$ ______________

20 Find θ correct to the nearest minute.

a $\tan\theta = \dfrac{3}{5}$ ______________

b $\cos\theta = 0.553$ ______________

c $\sin\theta = \dfrac{\sqrt{3}}{2}$ ______________

d θ 0.8 1.5 ______________

21 Draw a bearing of 200°.

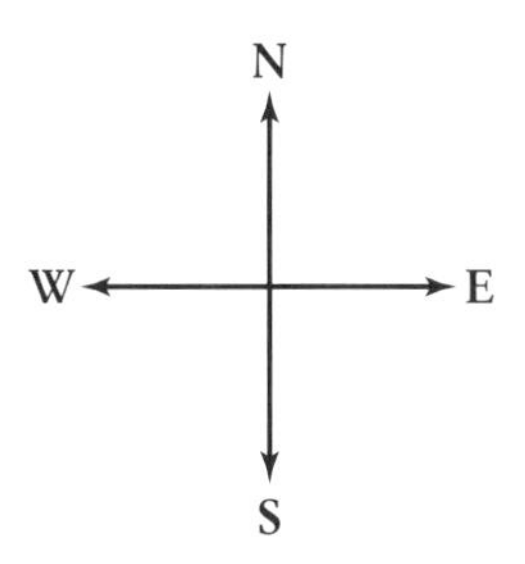

22 If $a = 6$, $b = 8$ and $C = 40^\circ$, evaluate to 2 decimal places:

a $\dfrac{1}{2}ab\sin C$ ______________

b $a^2 + b^2 - 2ab\cos C$ ______________

23 Find x correct to 2 decimal places.

a $x = \dfrac{25}{\tan 16^\circ 10'}$ ______________

b $x^2 = 3^2 + 9^2 - 2\times3\times9\cos 27^\circ$ ______________

____

c

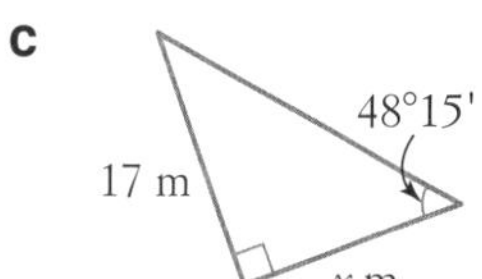

____

d

____

24 The angle of elevation from a point 550 m from the base of a tower to the top is 42°25′. Calculate the height of the tower to 2 decimal places.

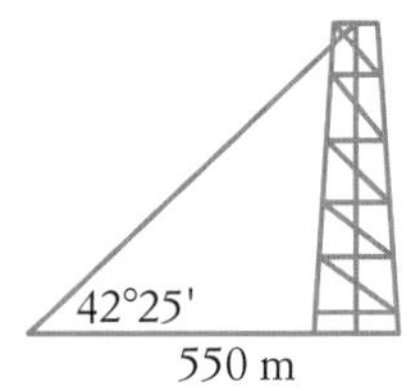

25 Find (to the nearest degree) the angle of depression of a boat that is 200 m from the base of a 74 m cliff.

26 From the lookout, a hiker walks north 8 km, then 6 km west to the camp.

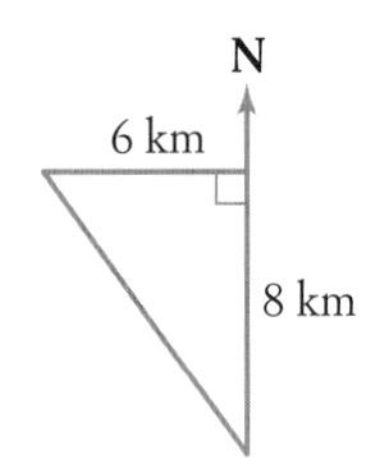

a How far is the camp from the lookout?

b What is the bearing (in degrees and minutes) of the camp from the lookout?

PART C: CHALLENGE

Bonus / 3 marks

Find an angle size θ that satisfies the equation $\sin\theta = \cos\theta$. Can you prove your answer?

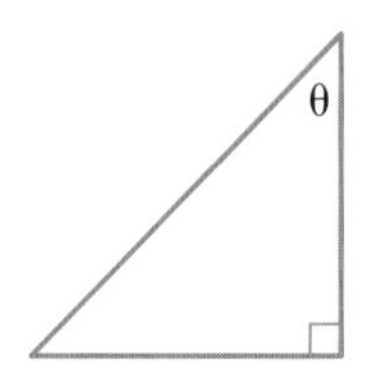

9780170454551

FINDING AN UNKNOWN ANGLE 15

YOU NEED TO USE THE SINE OR COSINE RULES TO SOLVE THESE

Find the size of the angle marked in each triangle, correct to the nearest degree.

1

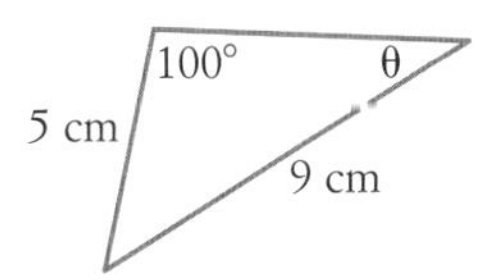

2

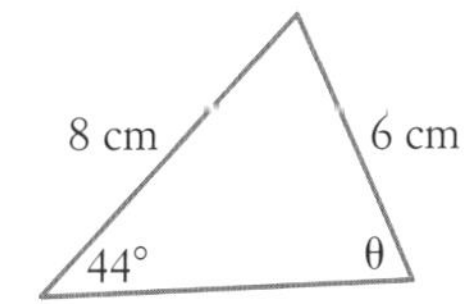

3

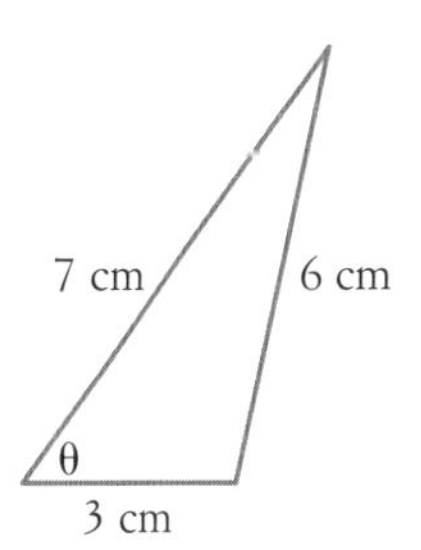

4

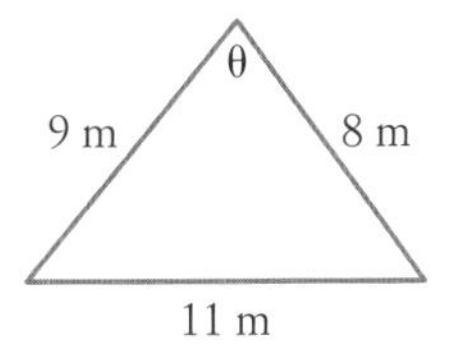

5 θ is obtuse

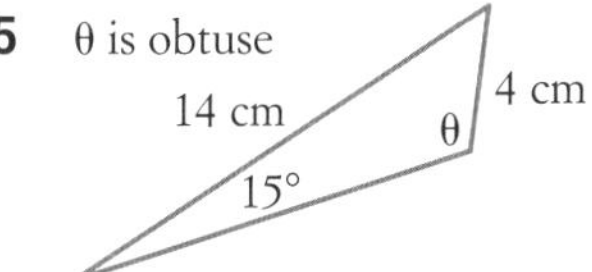

6

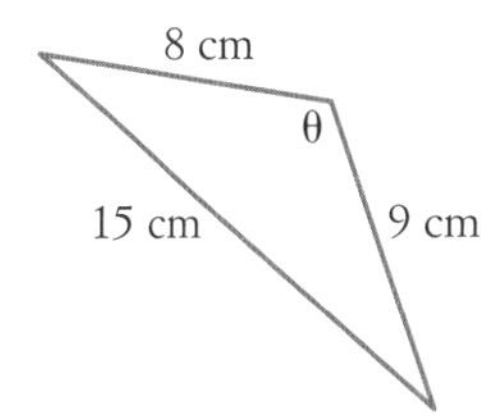

7

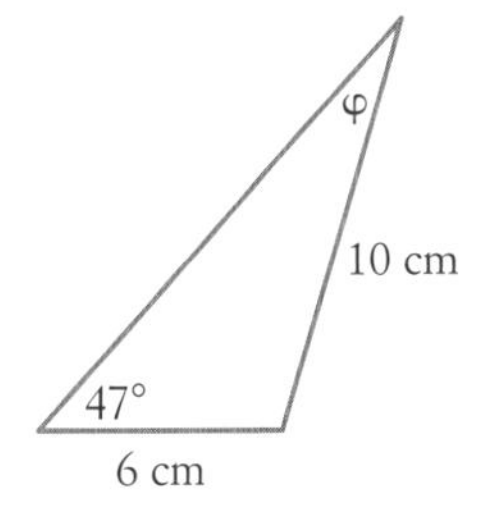

8

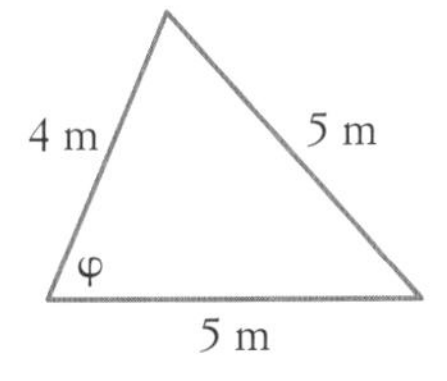

9

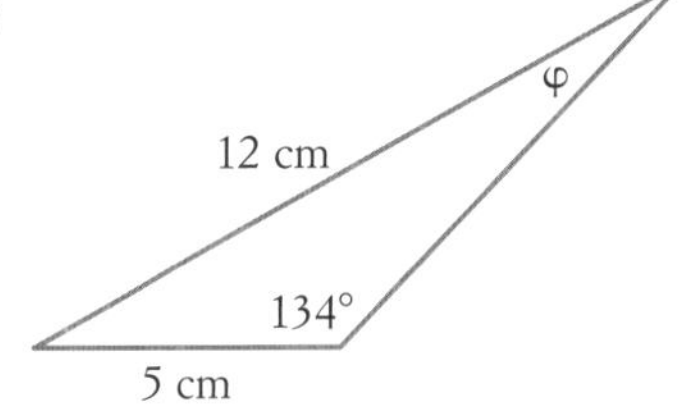

10

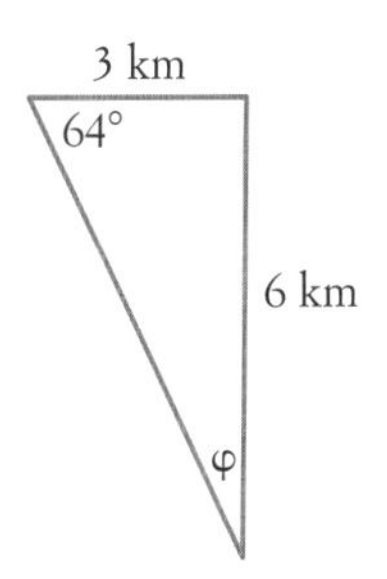

11 φ is obtuse

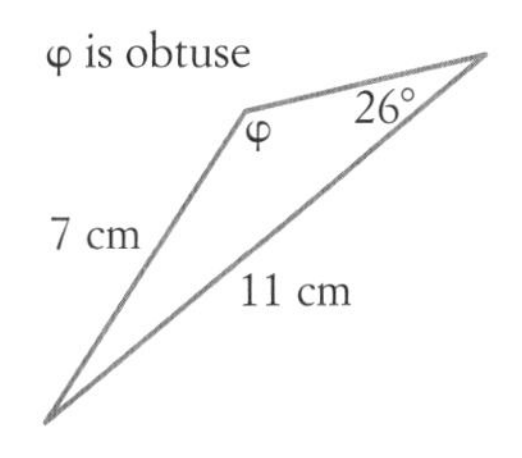

12

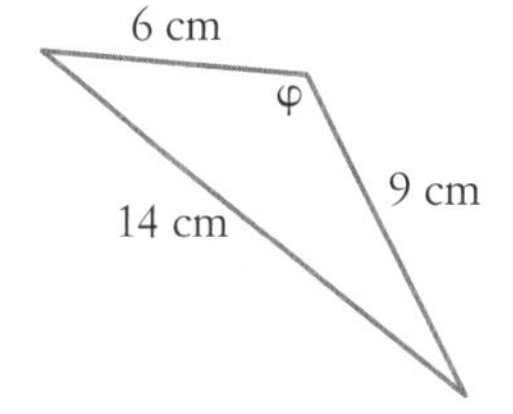

WS WORKSHEET

13

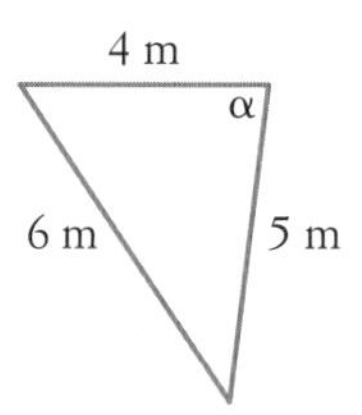

14

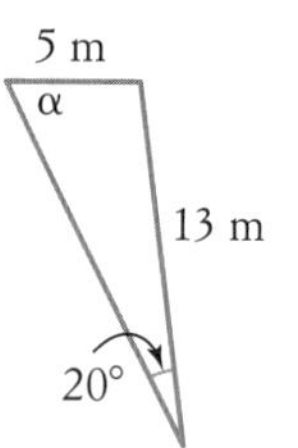

15

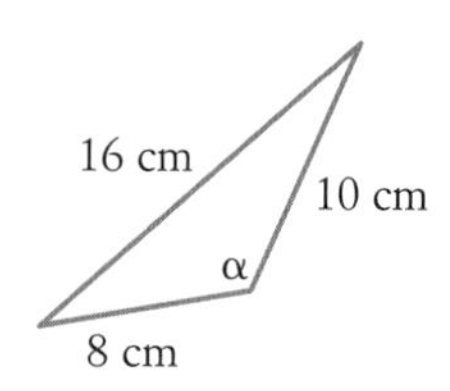

16 α is obtuse

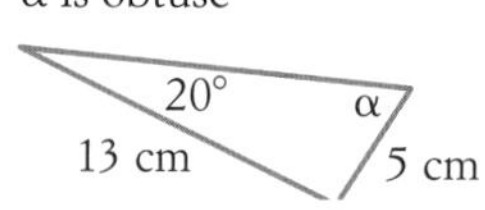

17

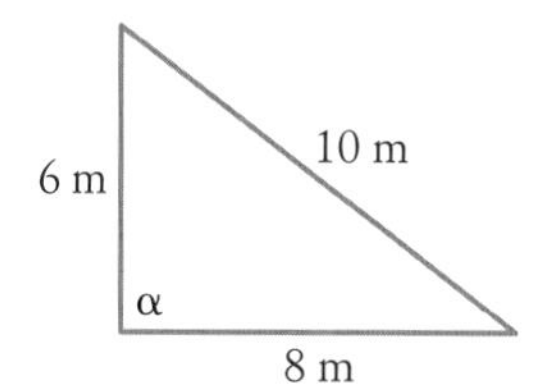

18

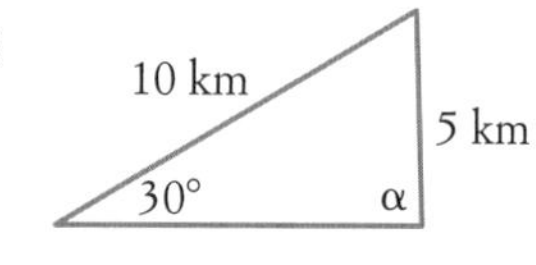

19

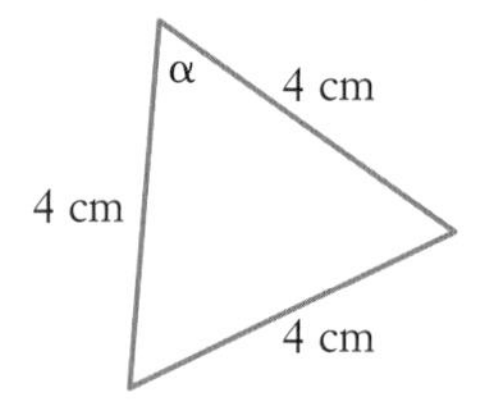

20

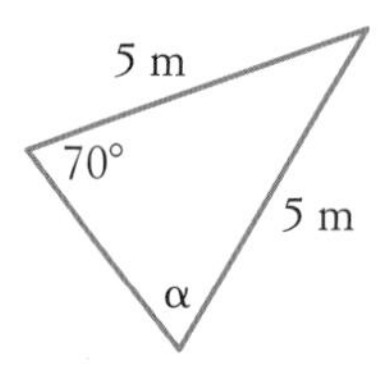

21

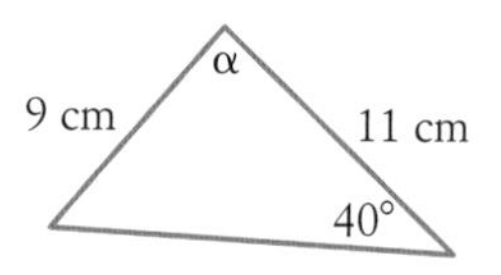

Mixed answers: 83°, 68°, 136°, 90°, 80°, 70°, 125°, 63°, 60°, 66°, 88°, 117°, 33°, 115°, 58°, 27°, 52°, 26°, 137°, 124°, 17°, 90°

FURTHER TRIGONOMETRY 1 (15)

Name:

Due date:

Parent's signature:

Part A	/ 8 marks
Part B	/ 8 marks
Part C	/ 8 marks
Part D	/ 8 marks
Total	/ 32 marks

HERE WE'LL LOOK AT THE TRIGONOMETRY OF ANGLES GREATER THAN 90°.

PART A: MENTAL MATHS

Calculators not allowed

1 Simplify $4\sqrt{18}+\sqrt{98}$.

2 Mai earns a salary of $78 000 p.a. How much does she earn each month?

3 Find the gradient of the line passing through points (4, 5) and (–6, 7).

4 Simplify.

a $\left(\dfrac{2m^4n^5}{m^5n^3}\right)^{-2}$

b $\left(9x^2\right)^{-\frac{3}{2}}$

5 If 3 coins are tossed together, what is the probability of tossing 1 head and 2 tails?

6 (2 marks) Solve $2x^2 - 9x - 5 = 0$.

PART B: REVIEW

1 For this rectangular prism, find:

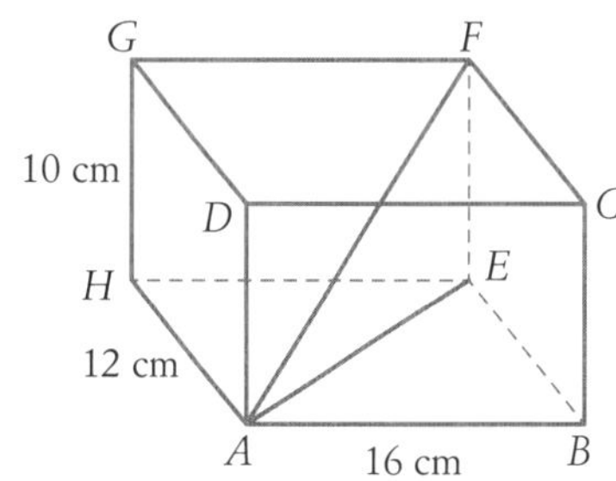

a AE

b AF, correct to one decimal place.

HW HOMEWORK

c $\angle FAE$, correct to the nearest degree.

2 The bearing of Braddon (B) 182 km from Ashville (A) is 138°.

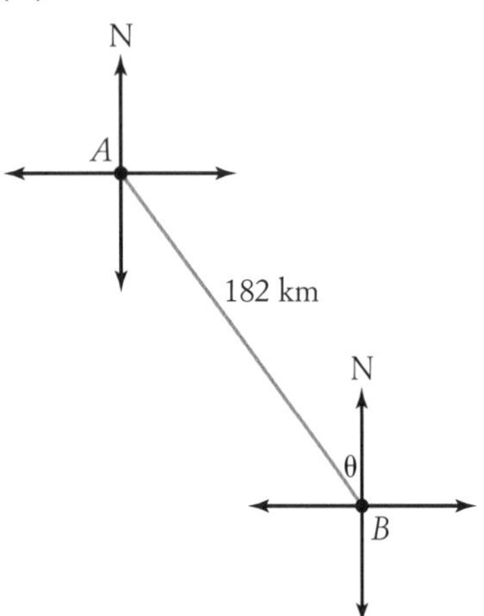

a Find the value of θ.

b How far south of Ashville is Braddon, to the nearest kilometre?

c What is the bearing of Ashville from Braddon?

3 Find:

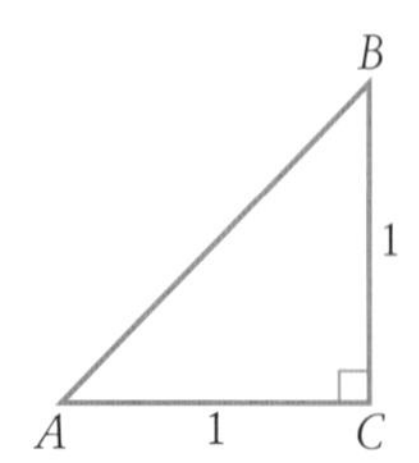

a the length of AB as a surd

b the size of angle A.

PART C: PRACTICE

› Trigonometry of angles greater than 90°
› Trigonometric equations

1 Evaluate tan 158° 22' correct to 2 decimal places.

2 If θ is acute, find θ if:

a $\sin 135° = \sin\theta$

b $\cos 100° = -\cos\theta$

c $\tan 153° = -\tan\theta$

3 (2 marks) Solve $\sin\theta = 0.74$ correct to the nearest degree, giving all possible acute and obtuse solutions.

4 (2 marks) Solve $\cos x = -\frac{3}{25}$, correct to the nearest minute, if x is obtuse.

9780170454551

PART D: NUMERACY AND LITERACY

1 This is the graph of $y = \cos\theta$.

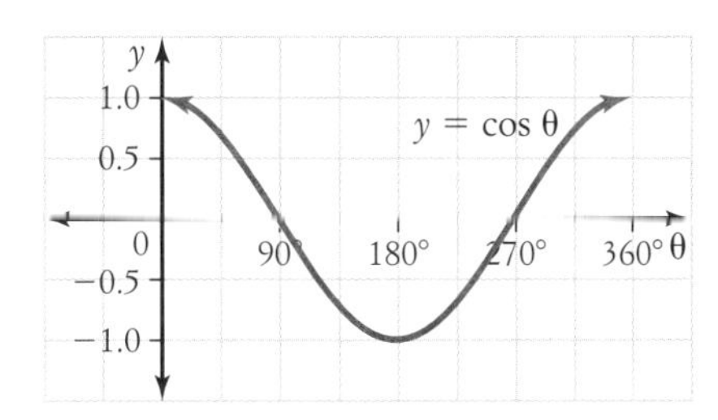

a If θ is obtuse, is $\cos\theta$ positive or negative?

b What is $\cos 90°$?

c What is the highest value of $y = \cos\theta$?

d For what values of θ is $\cos\theta = 0.5$?

2 $\triangle CBA$ is an equilateral triangle.

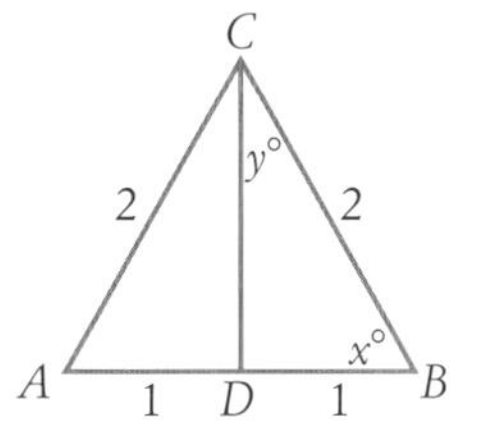

a Find its height CD as a surd.

b Find the values of x and y.

c Hence find the value of $\sin 60°$ as a surd.

C S F

HW HOMEWORK

15 FURTHER TRIGONOMETRY 2

Name:

Due date:

Parent's signature:

Part	Marks
Part A	/ 8 marks
Part B	/ 8 marks
Part C	/ 8 marks
Part D	/ 8 marks
Total	/ 32 marks

PART A: MENTAL MATHS

Calculators not allowed

1 Evaluate $125^{-\frac{2}{3}}$.

2 Rationalise the denominator of $\frac{18}{2\sqrt{2}}$.

3 (2 marks) Factorise $10\,000b^4 - 81a^4$.

4 (2 marks) Find the equation of the line perpendicular to $y = 5x - 2$ that passes through the point (1, 4).

5 For the graph of $y = 2x^2 - 8$, find:

a the y-intercept

b the x-intercepts

PART B: REVIEW

1 (3 marks) From the top of a 200 m tower, the angle of depression of a car is 50°.

a Draw a diagram to show this information.

b How far, correct to the nearest metre, is the car from the foot of the tower?

2 Which compass direction is:

a 270°?

b 135°?

3 (3 marks) A ship is due south of Adelaide. From the ship, on a bearing of 285°, a lighthouse is seen. If the lighthouse is 12 km due west of Adelaide, how far is the ship from the lighthouse, to one decimal place?

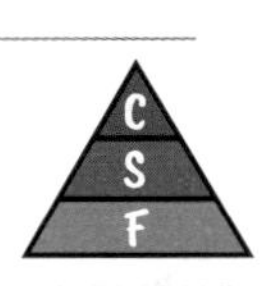

9780170454551

PART C: PRACTICE

› The sine and cosine rules

1 (4 marks) Find the value of each variable, correct to 2 decimal places.

a

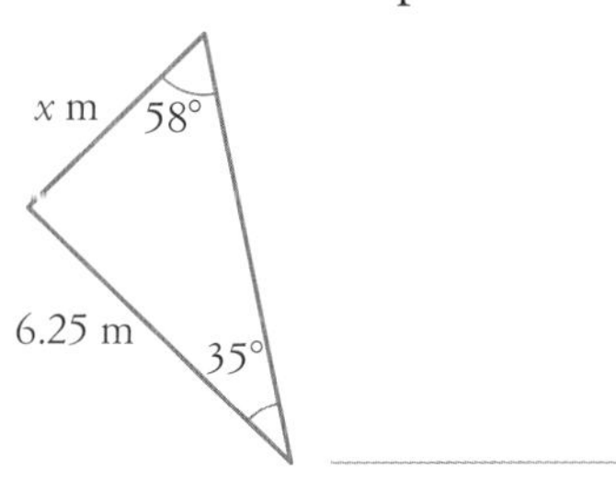

b

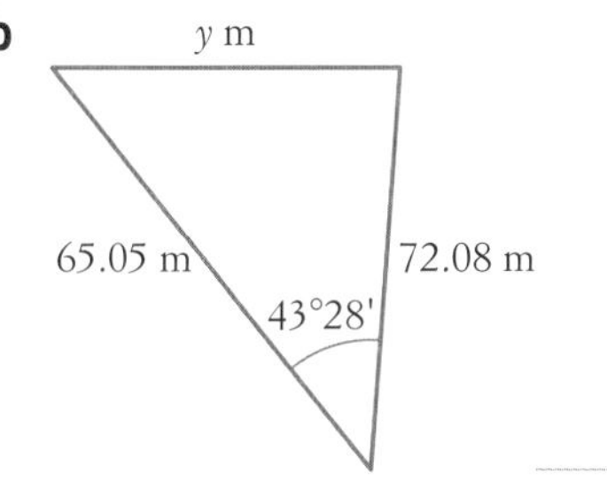

2 (4 marks) Find the value of each variable, correct to the nearest minute.

a

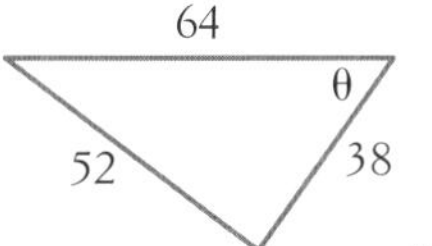

b

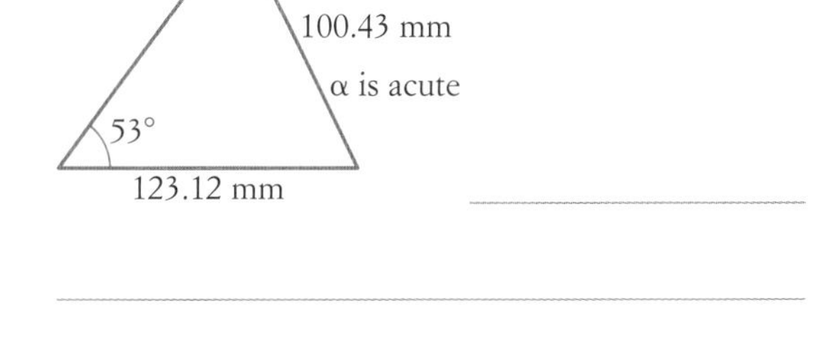

PART D: NUMERACY AND LITERACY

1 Which rule is a relationship between the 3 sides of a triangle and one of its angles?

____

2 Rewrite the cosine rule $c^2 = a^2 + b^2 - 2ab \cos C$ so that $\cos C$ is the subject of the formula.

____

3 Complete: In any triangle, the ratios of the sides to the sine of their __________ angles are __________ .

4 (2 marks) Find θ to the nearest degree if it is obtuse.

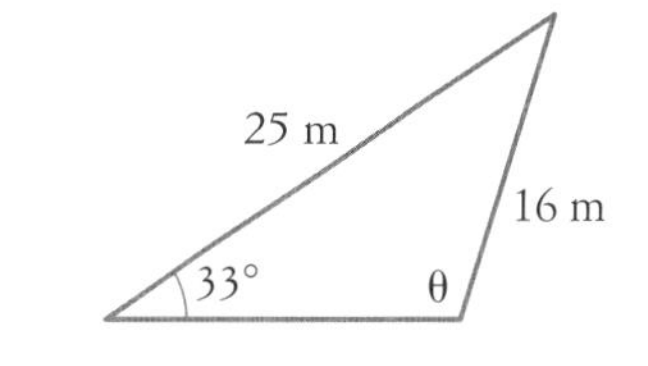

5 (3 marks) A plane flew on a bearing of 135° for 275 km. It then changed direction and flew another 316 km on a bearing of 223°. How far, correct to the nearest kilometre, is the plane from its starting point?

____

HW HOMEWORK

15 FURTHER TRIGONOMETRY 3

$A = \frac{1}{2}ab \sin C$ IS A FORMULA FOR THE AREA OF A TRIANGLE WHEN YOU KNOW THE LENGTHS OF 2 SIDES AND THE INCLUDED ANGLE BETWEEN THEM.

Name:

Due date:

Parent's signature:

Part A	/ 8 marks
Part B	/ 8 marks
Part C	/ 8 marks
Part D	/ 8 marks
Total	/ 32 marks

HW HOMEWORK

PART A: MENTAL MATHS

1 Find the interquartile range of each set of data.

a 4 7 7 9 11 12 13 14 17

b 0 1 2 3 4 5 6 7

2 (2 marks) Simplify $\dfrac{x^2 - 4}{2x^2 - 9x + 10}$.

3 (2 marks) Find, in terms of π, the surface area of a cylinder with base radius 5 m and height 10 m.

4 (2 marks) Sketch the graph of $y = 2^{-x}$, showing the y-intercept and the coordinates of one point.

PART B: REVIEW

1 (2 marks) Find the value of x, correct to the nearest whole number.

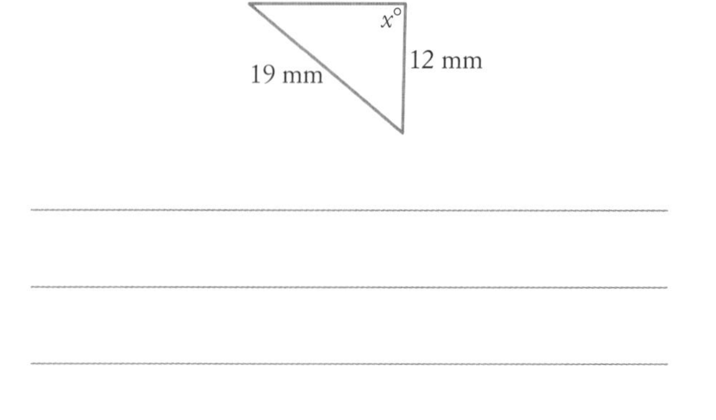

9780170454551

2 (4 marks) The angles of elevation of a building measured from 2 positions 30 m apart are 42° and 60°.

a Find $\angle ADB$.

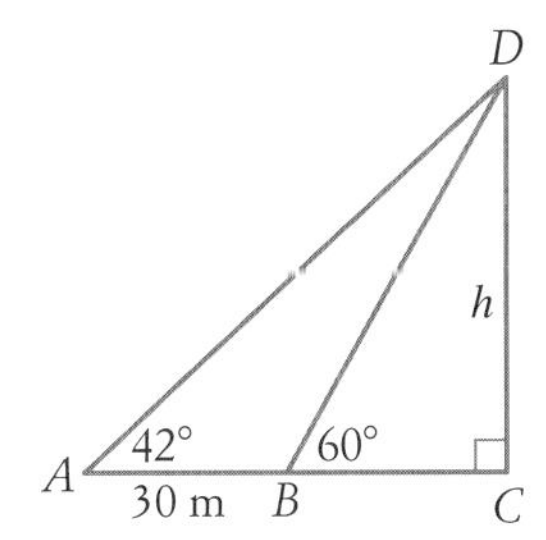

b Find, correct to 2 decimal places, the length of BD.

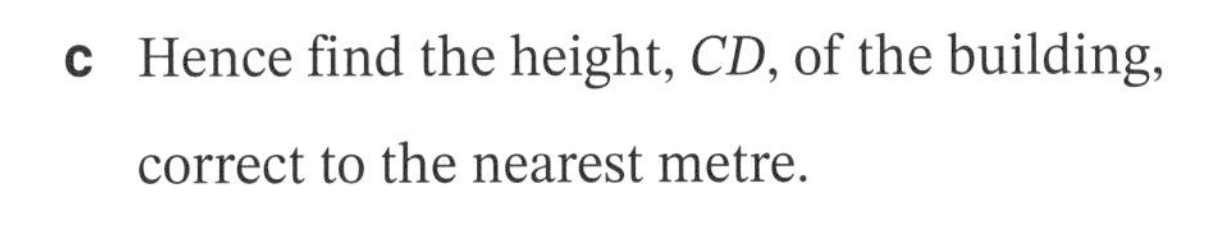

c Hence find the height, CD, of the building, correct to the nearest metre.

3 (2 marks) Find x, correct to one decimal place.

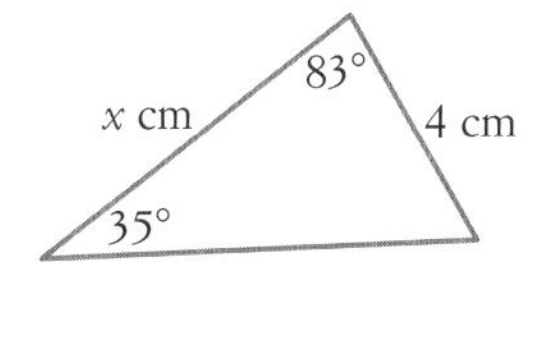

PART C: PRACTICE

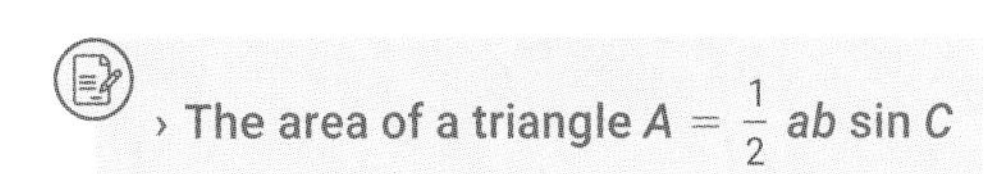

› The area of a triangle $A = \frac{1}{2}ab\sin C$

1 (5 marks) Find, correct to one decimal place, the area of each shape.

a

21.7 cm, 76°, 17.1 cm

b

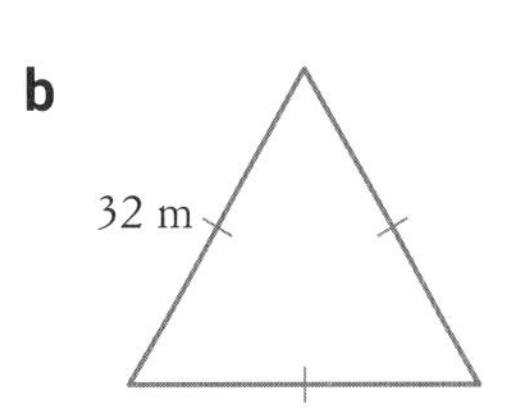

c

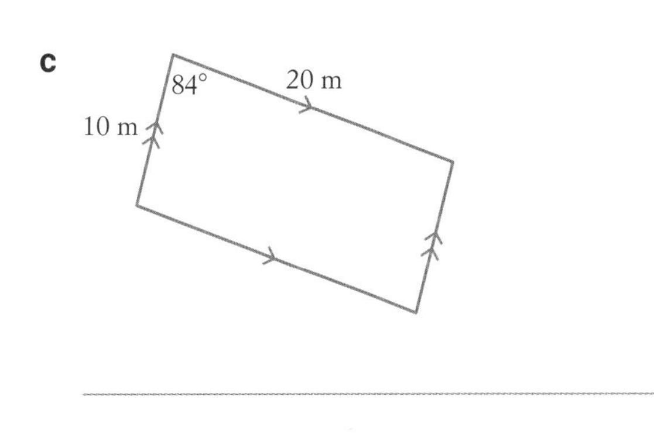

HW HOMEWORK

2 O is the centre of a circle of radius 35 cm.

A B 112° 35 cm O

Calculate, correct to one decimal place, the area of:

a sector OAB with angle 112°

____

b $\triangle OAB$

____

c the shaded segment

____

PART D: NUMERACY AND LITERACY

1 Complete: The area of a triangle with sides of length a and b and ____________ angle C is $A = \frac{1}{2}ab$ ____________.

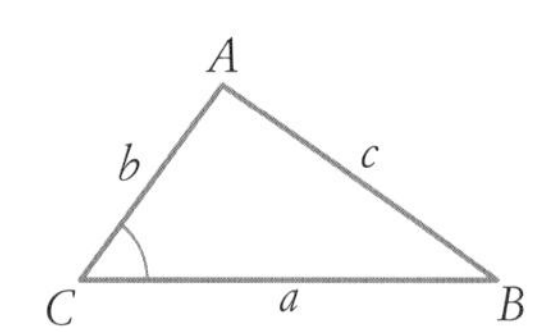

2 Which rule is used for triangle problems involving 2 sides and the 2 angles opposite them?

____

3 What is another name for the cosine rule if the angle used is 90°?

____

4 (4 marks) Ronan wants to run around this cross-country course.

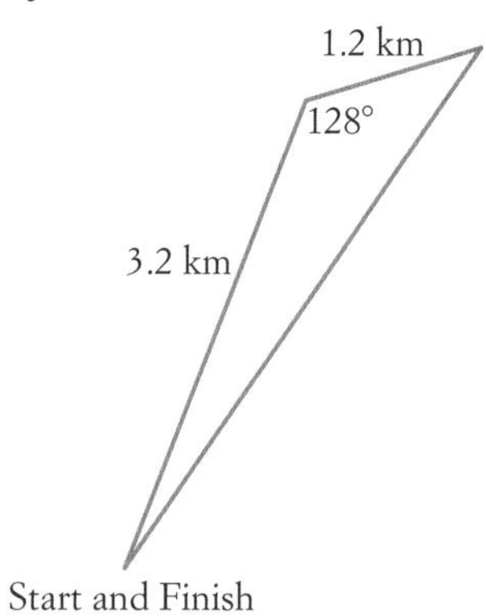

Calculate, correct to one decimal place:

a the area of the triangle

____

b the perimeter of the triangle

____

____

____

HW HOMEWORK

C S F

9780170454551

Chapter 1

StartUp assignment 1 PAGE 01

1 33.909
2 a $64x^6$ **b** $\frac{8}{27}$
3 $10d + 6$
4 Volume of a cylinder **5** $x = 14$
6 a 2 : 7 **b** 6 : 5
7 SAS **8** 5.4 L/hour
9 $\sqrt{20}$ or $2\sqrt{5}$ units **10** 65 cm^2
11 No **12** $r = 69$
13 1.6×10^{11}
14 a 30 **b** 130
c 3 **d** 16
e 0.75 **f** 1.75
15 a $\frac{3}{25}$ **b** $\frac{1}{8}$
16 a 0.09 **b** 0.065
c 0.015 **d** 0.008
17 23%
18 a $320 **b** $3.21
c $1725 **19** $5382
20 $864 **21** $144.45
22 $600 **23** $86.50
24 a $2100.15 **b** $3254.86
25 $756 **26** $2349
Challenge: At the end of 12 years

Compound interest PAGE 03

1 a 48 **b** 84
c 42 **d** 15
2 a 0.04 **b** 0.065
c 0.1325 **d** 0.0803
e 0.075 **f** 0.0375
g 0.006 **h** 0.001 48
3 a 0.011 25 **b** 0.015
c 0.002 625 **d** 0.009 225
4 a 0.000 411 0 **b** 0.000 205 5
c 0.000 591 8 **d** 0.000 458 1
5 a 0.045 **b** 0.0175
6 a $5832 **b** $6802.44 **c** $7346.64
7 a $4908.93 **b** $19 539.18
c $32 331.19 **d** $12 989.19
e $22 721.51 **f** $52 294.04
8 a $1049.91 **b** $13 928.37
c $6384.43 **d** $1084.20
9 a $2470.99 **b** $7166.79 **c** $12 860.09

10

	Principal	Rate (% p.a.)	Time	Compounded	Final amount	Interest
a	$5500	7%	4 years	Yearly	$7209.38	$1709.38
b	$6372.75	6.4%	6 years	Half-yearly	$9300	$2927.25
c	$20 000	12.6%	3.5 years	Monthly	$31 013.83	$11 013.83
d	$21 397.28	9%	2 years	Monthly	$25 600	$4202.72
e	$8194.17	14.8%	$2\frac{3}{4}$ years	Quarterly	$12 220	$4025.83

Depreciation PAGE 05

1 $28 343.52
2 a $13 384.40 **b** $20 615.60
3 $9686.44 **4** $198 766.99
5 a $277 992 **b** $216 833.76 **c** $61 158.24
6 a $8439.78 **b** 43.1%
7 a 90% **b** 81% **c** 59%
8 a $8162.10 **b** $6211.84 **c** 7 years
9 a $31 993.30 **b** $26 285.13 **c** 13 years
10 profit **11** 8 years
12 Disagree, the amount of depreciation changes each year because it is calculated from the previous year's value.

Interest and depreciation crossword PAGE 07

Down
1 repayment **4** investment
6 gross **7** wage
10 principal **11** loading
12 fortnightly **16** rate
18 annual **20** term
22 compound **23** final
25 leave **26** salary

Across
2 simple **3** flat
5 overtime **8** tax

9 deposit
13 commission
14 deduction
15 net
17 income
19 interest
21 per cent
23 formula
24 monthly
27 PAYG
28 quarterly
29 balance
30 piecework

Chapter 2

StartUp assignment 2

PAGE 11

1 $0.\dot{7}\dot{2}$
2 \$1085
3 1000
4 irrational
5 2.60
6 35.1 L
7 $SA = 2\pi r^2 + 2\pi rh$
8 a 4.105×10^{-2} **b** 4
9 $27y^2 - 3$
10 360°
11 $25u^2 - 40u + 16$
12 $\frac{2m}{5}$
13 $d = 3$
14 $x = 16$
15 $-13, -9, -5, -1$
16 $-\frac{1}{3}$
17 (5, 0)
18 24 units^2
19 a 10 units **b** (1, 3) **c** $\frac{3}{4}$

20

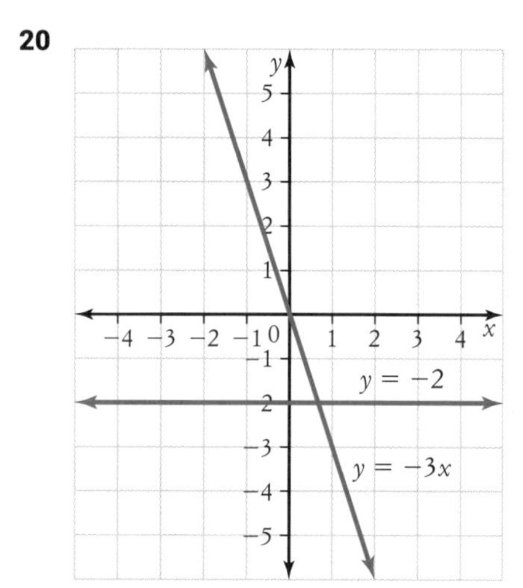

21 a one pair of opposite sides parallel
b all sides equal
c opposite sides equal and parallel
22 $3\sqrt{5}$
23 $y = -4$
24 Sample answer only. Other (similar) negative gradient answers are possible. Teacher to check.

25

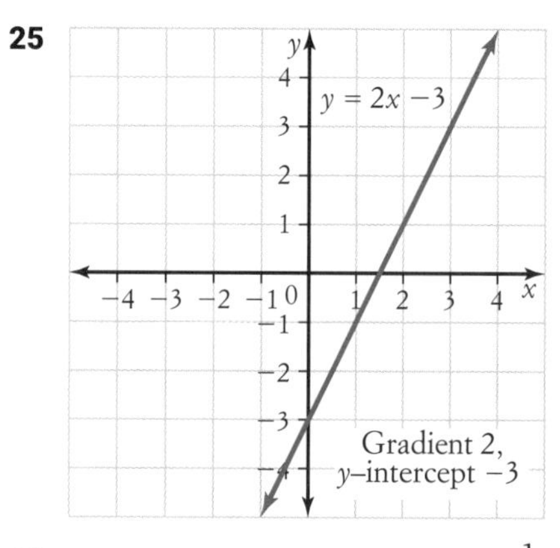

26 a 2 **b** $5\frac{1}{2}$
c 2 **d** 4
27 Yes
28 Teacher to check
Challenge: 12, circle

Linear equations code puzzle

PAGE 14

1 Y
2 O
3 B
4 S
5 F
6 W
7 A
8 L
9 R
10 T
11 E
12 H
13 X
14 D
15 I
16 J
17 N
18 U
19 Z
20 C
21 M
22 G
23 V
24 K
25 P

Answer: What did one parallel line say to another at the party?
'Excuse me, you look very familiar but we've never met, have we?'

Writing equations of lines

PAGE 16

1 Teacher to check
2 Any equation in the form $y = 6x + c$
3 Any equation in the form $y = c$
4 Any equation in the form $y = -mx$
5 Any equation in the form $y = \frac{1}{3}x + c$
6 Any equation in the form $y = mx + 5$
7 Any equation in the form $y = \frac{1}{6}x + c$
8 Any equation in the form $y = -\frac{2}{3}x + c$
9 $x = -1$
10 Any equation in the form $y = mx + c$, with $m > 2$
11 Teacher to check
12 Teacher to check
13 Any equation in the form $y = 2x + c$
14 Any equation in the form $y = c$

9780170454551

15 Any equation in the form $y = mx + 5$

16 $y = -5$

17 Any equation in the form $y = mx - 5$

18 $y = -5x + 10$

19 $y = \frac{3}{10}x - 3$

20 Any equation in the form $y = mx + c$, with $m > -\frac{1}{2}$

21 $y = 3x - 7$

22 Any equation in the form $y = -\frac{3}{2}x - c$

23 $y = \frac{4}{9}x - 4$

24 $y = -x - 4$

Graphing lines crossword PAGE 17

Across

3 steepness
6 intercept
8 origin
10 graph
13 line
14 general
17 equation
18 constant
21 reciprocal
23 surd
24 hypotenuse
26 negative
27 positive
28 vertical
29 exact

Down

1 form
2 length
4 plane
5 Cartesian
7 gradient
9 Pythagoras
11 run
12 interval
15 rise
16 midpoint
18 coordinates
19 axes
20 parallel
22 inclination
25 linear
27 point

Chapter 3

StartUp assignment 3 PAGE 23

1 Example answer given. Other triangles with 2 equal sides and an included obtuse angle are possible.Teacher to check.

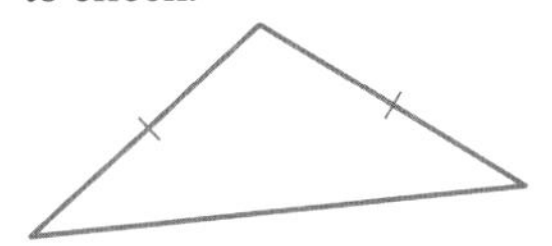

2 $\frac{1}{25k^2}$

3 a 13.25 **b** 12

4 $6ab$

5 115°

6 85%

7 3.78×10^7

8 $x = 7$

9 a 8.94 units **b** -2

10 $\frac{5b}{a}$

11 $0.8\dot{3}$

12 7%

13 10

14 a

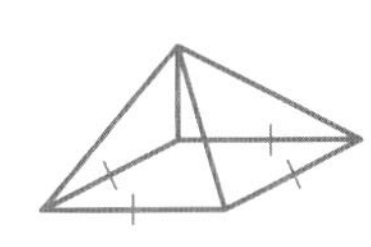

b

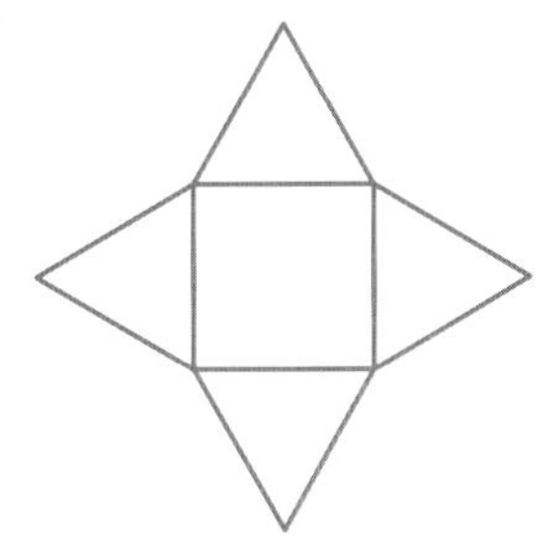

15 a 37.70 cm **b** 113.10 cm²

16 a 384 cm² **b** 512 cm³

17 a 10 000 **b** 1000

18 a triangular prism **b** 5 **c** rectangle, triangle **d** 45 cm³

19 a 84.82 **b** 113.10

20 $A = \frac{1}{2}(a + b)h$

21 a rhombus **b** 38.5 cm²

22 220 cm

23 7 cm

24 A prism has a uniform cross-section; the side faces of a pyramid meet at a point.

25 8 m

26 a 15 cm² **b** 12 cm² **c** 13.96 cm²

27 6 cm

Challenge: $\frac{2}{3}$

Back-to-front problems PAGE 26

1 $l = 7$ cm
2 $h = 2$ cm
3 $h = 4$ cm
4 $s = 7$ cm
5 $s = 4.5$ cm
6 $h = 12$ m
7 $r = 4$ cm
8 $w = 6$ cm
9 $h = 3$ m
10 $h = 14$ cm
11 $s = 3.8$ cm
12 $h = 15$ mm
13 $h = 25$ cm
14 $r = 5$ cm
15 $h = 20$ cm

Surface area and volume crossword PAGE 28

Across

3 square
4 annulus
5 area
8 cross section
11 triangular
12 sphere
16 semicircle
19 kilolitre
20 quadrant
22 circumference
23 diameter
27 open
28 external
29 radius
30 Pythagoras
32 pyramid
33 length

Down

1 capacity
2 cylinder
6 circle
7 cone
9 surface
10 sector
12 solid
13 rectangular
14 prism
15 closed
17 litre
18 base
21 pi
24 formula

25 height **26** net
31 width

Surface area

PAGE 30

1 2950 cm^2 **2** 1782 mm^2
3 1.5 m^2 **4** 408 cm^2
5 70.7 m^2 **6** 111.9 cm^2
7 6283.2 mm^2 **8** 1350 mm^2
9 51 cm^2 **10** 3240 mm^2
11 879.6 cm^2 **12** 311.04 m^2

Chapter 4

StartUp assignment 4

PAGE 37

1 -4 **2** \$3300, \$1800
3 $m^2 + 6m$ **4** $135°$
5 7.46×10^{-6} **6** 6
7 false, but a square is a rhombus
8 $\frac{1}{4}$ **9** 376.99 cm^2
10 $x = 16\frac{1}{2}$ **11** 126
12

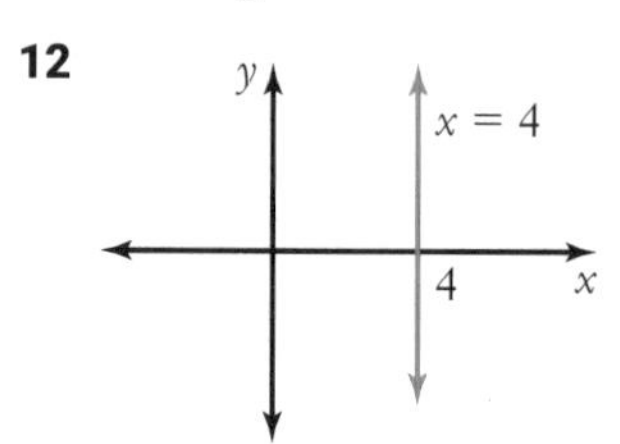

13 \$45.00
14 a $23°$ **b** 30 cm^2
15 a 1, 3, 9 **b** 1, 3, 17, 51
16 a $21m^3$ **b** $5x$ **c** $12k^4$
d $-8g^2$ **e** $16pq$ **f** $9u^8$
17 a 4 **b** $8y$
18 1, 4, 9, 16, 25, 36, 49, 64, 81, 100
19 a $3x + 21$ **b** $10m - 2n$ **c** $4 - 12g$
d $-12p - 10$ **e** $m^2 + 7m$ **f** $3p^2 - 4p$
g $6k^2 - 12k$ **h** $-15y + 5y^2$
20 a $3(x + 4)$ **b** $10(m - 2)$ **c** $p(3p - 5)$
d $-2(4m - 9)$ **e** $-2y(7y + 4)$

Challenge: Numerous solutions.

Balance 6 coins and 6 coins to find which side holds the counterfeit. Then balance 3 coins and 3 coins from the side identified as holding the counterfeit, to again find which side holds the counterfeit. Then choose 2 coins and balance 1 and 1. If they balance, then the 3rd coin is the counterfeit. If they don't, then the lighter one is the counterfeit.

Algebraic fractions

PAGE 39

1 $\frac{3x}{5}$ **2** y
3 $\frac{7d}{4}$ **4** $\frac{4h}{3}$
5 $\frac{23h}{15}$ **6** $\frac{53k}{14}$
7 $\frac{47x}{24}$ **8** $\frac{5}{x}$
9 $\frac{32}{3d}$ **10** $\frac{3y}{2}$
11 $2p$ **12** r
13 $\frac{2e}{5}$ **14** $\frac{17d}{12}$
15 $\frac{20p}{21}$ **16** $\frac{-36}{5d}$
17 r **18** $\frac{-g}{4}$
19 $\frac{ey}{3}$ **20** $\frac{8rt}{3}$
21 $45y$ **22** $20g$
23 $10f$ **24** $\frac{10hg}{3}$
25 $\frac{11d}{6}$ **26** $\frac{3r^2}{5}$
27 $\frac{25e^2}{49d^2}$ **28** $\frac{3d}{10}$
29 $\frac{5}{2}$ **30** $\frac{21ef}{4}$
31 $\frac{16a}{9c}$ **32** $4y^2$
33 $\frac{h}{2g^2}$ **34** $\frac{21d}{10}$
35 $\frac{9mn}{5}$ **36** $\frac{7h}{24f}$

Algebra crossword

PAGE 40

Across

3 product **8** binomial
10 numerator **13** brackets
14 indices **16** index
18 base **19** factors

Down

1 constant **2** quadratic
3 power **4** term
5 factorise **6** highest
7 coefficient **9** algebraic
11 reciprocal **12** HCF
15 expand **17** denominator

Mixed expansions

PAGE 50

1 $b^2 + 12b + 35$ **2** $2b^2 - 36b$
3 $9b^2 + 6b + 1$ **4** $b^2 - 64$
5 $b^2 + 6b - 27$ **6** $b^2 - 1$
7 $b^2 - 9$ **8** $bc - 2b - c^2 + 2c$
9 $b^2 + b + bc + c$ **10** $b^2 + 16b + 64$
11 $x^2 - 12x + 32$ **12** $x^2 - 14x + 49$
13 $4x^2 - 3x - 10$ **14** $4x^2 + 8x - 45$
15 $x^2 - 10x + 25$ **16** $-14x + 21$

9780170454551

17 $36x^2 - 4$

18 $4x^2 - 24x + 36$

19 $4x^2 - 12x - 7$

20 $4x^2 - 9$

21 $-m^2 - 7m$

22 $m^2 + 14m + 49$

23 $2m^2 - 10m - 28$

24 $6m^2 + m - 12$

25 $16m^2 - 1$

26 $m^2 + 7m + 12$

27 $m^2 + 4m - 5$

28 $4m + 4$

29 $m^2 + 3m - 10$

30 $9m^2 - 12m + 4$

Chapter 5

StartUp assignment 5 PAGE 51

1 $125a^3$

2 B

3 6

4 $3 - 2y$

5 trapezoidal prism

6 **a** 169.65 cm³ **b** 169.65 cm²

7 $\frac{5}{8}$

8 $k = 10$

9 36

10 $d = 26\frac{2}{3}$

11 $y = 0$

12 11.5%

13 length 125 cm, width 100 cm

14 560 km

15 **a** 19 **b** 2 **c** 4

d 2.42 **e** 2 

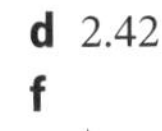

f

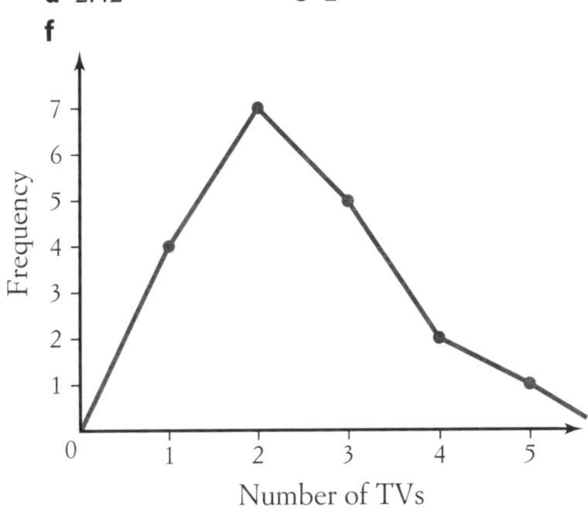

16 A census surveys the whole population; a sample surveys a part of the population.

17 **a** 64 **b** 24 **c** 64

d 11 **e** 64.62

18 **a**

Score	f	cf
1	2	2
2	5	7
3	6	13
4	8	21
5	5	26
6	2	28
7	1	29
8	1	30

b 4 **c** 4

d 7 **e** 13

Challenge: 63 matches, 6 rounds

Interquartile range PAGE 53

	Median	Q_1	Q_3	Interquartile range
1	15	11	20	9
2	18	13	$25\frac{1}{2}$	$12\frac{1}{2}$
3	11	5	14	9
4	4	1	9	8
5	$22\frac{1}{2}$	20	31	11
6	30	19	40	21
7	64	57	68	11
8	3	2	5	3
9	85	83	88	5
10	30	15	45	30
11	4	3	6	3
12	2	1	3	2
13	7	$6\frac{1}{2}$	$8\frac{1}{2}$	2
14	9	9	9	0
15	9	8	10	2
16	8	7	9	2
17	33	$23\frac{1}{2}$	$36\frac{1}{2}$	13
18	2	1	4	3

Box-and-whisker plots PAGE 54

1 **a** **i** drivers **ii** passengers

iii drivers **iv** drivers

b 350 **c** 80

d 80 **e** negatively

f 80

g consistent, smallest spread

2 **a**

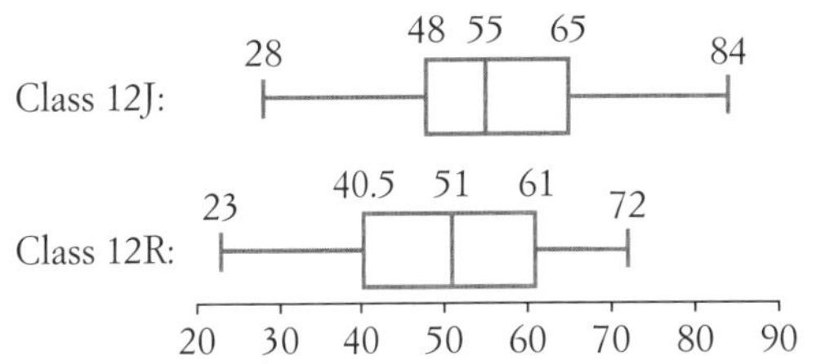

b Class 12J did better, that class had higher scores displayed in the box-and-whisker plot.

c Class 12J has the greater range; but class 12R has the greater interquartile range, and the greater range if the outlier in 12J is excluded.

d Class 12J; 28

e The distribution for class 12J is symmetrical and the distribution for class 12R is negatively skewed.

f False. Even if only the bottom 25% of each class is considered, class 12J still outperforms class 12R.

Data crossword PAGE 56

Across

2 quartiles
6 histogram
7 scatter
8 stem
11 bias
13 spread
14 order
15 five number
18 whisker
20 median
21 interquartile
22 mean
24 location
26 strong
27 bivariate
28 mode

Down

1 data
3 lower
4 plot
5 distribution
7 symmetrical
9 upper
10 range
12 skewed
15 frequency
16 middle
17 fifty
19 cumulative
23 cluster
25 outlier

Chapter 6

StartUp assignment 6 PAGE 66

1 $\frac{1}{9}$
2 $20\frac{5}{6}$ m/s
3 A
4 $\equiv$
5 $83\frac{1}{3}\%$
6 $p = 78$
7 2 km
8 a $\sqrt{41}$ **b** $\frac{4}{5}$
9 1 050 000
10 mode
11 9 : 6 : 2
12 a 144 cm^2 **b** 112 cm^3
13 6×10^{-3}
14 $p = 4\frac{2}{3}$
15 false
16 $y = -7$
17 a $10a - 6$ **b** $-12a + 24$
18 Yes
19 $k = 16$
20 $r = -3$
21 20, 18, 16, 14, 12, 10
22 $y = 6$
23 $m = 8\frac{1}{3}$
24 a $14r + 4$ **b** $t^2 + 8t - 9$
25 $p = 33$
26 $h = 9\frac{1}{2}$
27 37, 39, 41

28

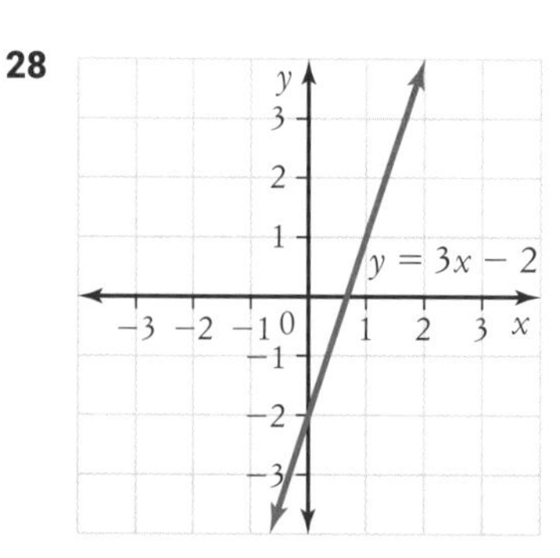

29 110°
30 $x = 3$
31 a \$240 **b** 3 hours
32 Yes
33 $x = 7$

Challenge: 83 emus, 167 pigs

Graphing inequalities PAGE 68

1

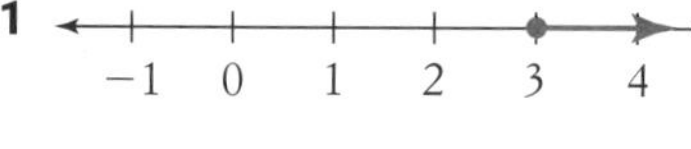

2 −1 0 1 2 3 4

3 −4 −3 −2 −1 0 1

4 1 2 3 4 5 6

5 −3 −2 −1 0 1 2

6 −3 −2 −1 0 1

7 −3 −2 −1 0 1

8 1 2 3 3.5 4 5 6

9 1 2 3 4 5

10 −1.5
−3 −2 −1 0 1 2

11 −2 −1 0 1 2

12 −4 −3 −2 −1 0 1

13 −2 −1 0 1 2 3

14 2 3 4 5 6 7

15 0 0.5 1 1.5 2

16 −7 −6 −5 −4 −3 −2

17 −3 −2.5 −2 −1.5 −1 −0.5

9780170454551

18 0 1 2 3 4 5

19 2 4 6 8 10 12

20 −5 −4 −3 −2 −1 0

21 −1 $-\frac{1}{2}$ 0 $\frac{1}{2}$ 1 $1\frac{1}{2}$

22 5 $5\frac{1}{2}$ 6 $6\frac{1}{2}$ 7 $7\frac{1}{2}$

23 −7 −6 −5 −4 −3 −2 −1

24 2 $2\frac{1}{2}$ 3 $3\frac{1}{2}$ 4 $4\frac{1}{2}$

Equations and inequalities crossword PAGE 70

Down

1 factorise
2 less
5 brackets
9 line
10 solution
11 negative
14 two
15 quadratic
18 LCM
21 formula

Across

3 subject
4 substitute
6 expand
7 greater
8 inequality
12 solve
13 equation
16 root
17 variable
19 inverse
20 multiple
21 fraction
22 RHS
23 check
24 LHS

Equations review PAGE 78

1 $y = 1$
2 $a = -9\frac{1}{3}$
3 $x = 5$
4 $k = 23$
5 $d = -\frac{1}{4}$
6 $x = \pm 6$
7 $n = \pm 3$
8 $m = 2$
9 $a = -4$
10 $p = 16$
11 $b = 27\frac{1}{2}$
12 $c = 4\frac{1}{11}$
13 $e = -2\frac{1}{2}$
14 $y = \pm 1$
15 $g = \pm 5$
16 $r = 24\frac{1}{2}$
17 $m = 4\frac{1}{3}$
18 $b = \pm 8$
19 $n = 10$
20 $w = -2\frac{7}{12}$

Chapter 7

StartUp assignment 7 PAGE 79

1 1.39
2 $\frac{1}{8}$
3 40 000
4 198
5 143.35
6 $5y$
7 3770 cm^3
8 6150
9 360°
10 108
11 $\frac{3}{10}$
12 56°
13 \$1300, \$650, \$2600
14 0
15 \$456.50
16 **a** −8, −2, 1, 7
b

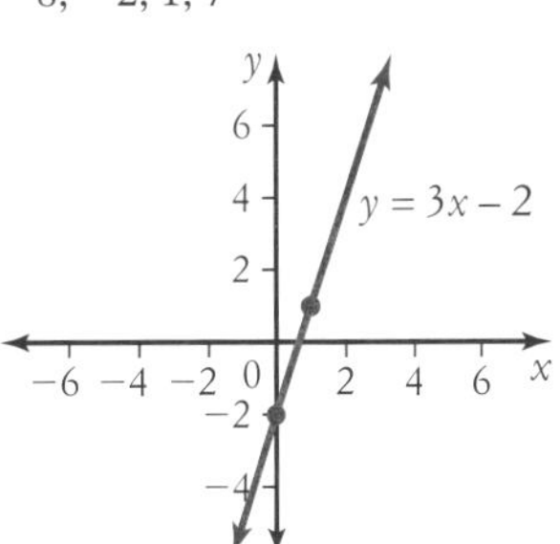

c 3 **d** −2 **e** $\frac{2}{3}$

17 0, −4, −3, 5
18 **a** $x = -2$ **b** $y = 18$
19 Example answer only.
Other similar answers possible.
Teacher to check.
20 **a** 12 **b** 48 **c** 3
21 **a** $y = 4x + 3$ **b** $y = x^2$
22 **a** 60 km/h **b** 12:30 p.m. **c** 150 km
23 **a** 1 **b** $\frac{1}{2}$
c $y = \frac{1}{2}x + 1$
24 **a** $-1\frac{1}{3}$ **b** $\frac{2}{5}$ **c** 8

Challenge: 16 m

Graphs crossword PAGE 84

Across

3 line
6 asymptote
7 graph
9 point
10 circle
11 centre
13 direct
18 coordinates
19 vertex
22 constant
24 variable
25 horizontal
26 axis

Down

1 equation
2 concave
4 exponential
5 quadratic
8 coefficient
11 curve
12 conversion
14 inverse
15 vertical
16 origin
17 proportion
20 radius
21 intercept
23 parabola

Chapter 8

StartUp assignment 8 PAGE 88

1 $\frac{5}{7}$
2 A
3 $x = -12$
4 $72
5 10 000
6 $12\frac{1}{2}$%
7 −1
8 **a** $6x - 6$ **b** $2x^2 - 6x$
9 120°
10 m^4
11 $9520
12 150 cm³
13 9 : 5
14 104
15 19 min
16 270°
17 **a** 153 **b** 34
18 **a** 19.6 **b** 81 **c** 18
19 **a** m **b** w
20 7 h 9 min
21 **a** In both triangles, $x = 30$ **b** 10 units
22 57
23 18
24 **a** $x = 35$ **b** $h = 20$ **c** $d = 5$ **d** $y = 2\frac{1}{2}$
25 $(90 - a)°$
26 $x = 1, y = 30°$
27 **a** 12 **b** $\angle T$
28 **a** 1 cm **b** 1.41 cm

Challenge: $\theta = 45°$

Finding an unknown angle PAGE 90

1 34°
2 61°
3 39°
4 48°
5 54°
6 41°
7 59°
8 45°
9 57°
10 54°
11 61°
12 39°
13 36°
14 63°
15 37°
16 73°
17 27°
18 50°
19 37°
20 36°
21 11°

Elevations and bearings PAGE 92

1 9.51 km
2 117 cm
3 104 m
4 2264 m
5 2.23 km
6 1581 m
7 254.76 nm
8 316°
9 33°45′
10 12.68°
11 160 nm
12 **a** Strawberry Field (4.77 m)
b Macarthur Park (6.53 m)

Trigonometry crossword PAGE 94

Across

1 minute
3 three
7 ratio
11 angle
13 bearing
15 trigonometry
17 opposite
20 adjacent
22 right angled
25 northeast
26 horizontal
28 southwest
29 theta
30 sine
31 depression
32 southeast

Down

2 elevation
4 hypotenuse
5 tangent
6 figure
8 inverse
9 alternate
10 westsouthwest
12 mnemonic
14 denominator
16 degree
18 east
19 clinometer
21 northwest
23 north
24 cosine
27 west

A page of bearings PAGE 101

1 130°
2 280°
3 240°
4 045°
5 305°
6 230°
7 110°
8 060°
9 120°
10 350°
11 255°
12 020°
13 155°
14 210°
15 290°
16 135°
17 310°
18 015°
19 105°
20 190°

Chapter 9

StartUp assignment 9 PAGE 102

1 12.2
2 $3280.77
3 $n + 1$
4 area of a rhombus or kite
5 $9.1^2 = 3.5^2 + 8.4^2$
6 **a** 64 **b** $\frac{1}{9}$
7 **a** $2x + 18$ **b** $\frac{p^2}{5}$
8 6.28 cm²
9 0.000 028
10 $x = 2\frac{1}{2}$
11 SSS, SAS, AAS, RHS
12 $x = 19$
13 502.65 cm²
14 **a** $-40g + 32$ **b** $13y - 30$ **c** $-18a - 6$ **d** $75b - 67$
15 **a** $p = -5$ **b** $f = 8.5$ **c** $a = -1.5$ **d** $g = 19.2$ **e** $x = 13$ **f** $a = 3\frac{2}{3}$

9780170454551

16 **a** $y = 29$ **b** $y = 44$ **c** $y = 7$

17 **a** $a = -26$ **b** $a = 90$ **c** $a = 35$

18 **a** yes **b** yes **c** no

19

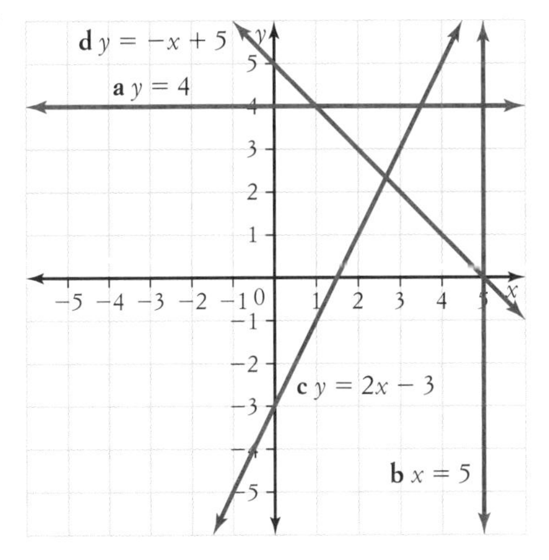

20 74 and 76

21 33, 35, 37

Challenge: Timmy

Intersection of lines PAGE 104

1 (3, 5)

2 (1, 0)

3 (4, –2)

4 (–2, 6)

Simultaneous equations crossword PAGE 105

Across

7 solve **11** intersection **12** point

14 satisfy **15** substitution **16** method

17 solution

Down

1 simultaneous **2** problem **3** variable

4 algebraic **5** axes **6** graphical

8 equations **9** elimination **10** linear

13 coefficient

Chapter 10

StartUp assignment 10 PAGE 111

1 1 000 000 **2** $\frac{11a}{15}$

3 Equal or bisect each other **4** \$80 and \$32

5 376.33 cm^3 **6** \$916

7 $y = 36$ **8** 3

9 70 **10** $d = 7$

11 \$24.50 **12** \$10 200

13 30

14 **a** 2.4 m **b** 34.56 m^2

15 58%

16 **a** $\frac{7}{20}$ **b** $\frac{1}{5}$ **c** $\frac{1}{4}$

17 **a** $\frac{1}{6}$ **b** $\frac{17}{24}$ **c** $\frac{5}{6}$

18 $\frac{1}{6}$ **19** Teacher to check

20 **a** 0.55 **b** $\frac{1}{6}$

21 Impossible, it cannot happen

22 0.5

23 Reds win, Blues win, or a draw

24 $\frac{2}{5}$

25 0.02 **26** $\frac{5}{8}$

27 55%

28 **a** $\frac{1}{4}$ **b** $\frac{1}{13}$

c $\frac{3}{13}$ **d** $\frac{5}{13}$

29 19.2%

Challenge: 120

Probability crossword PAGE 113

Across

3 table **6** without

7 replacement **8** conditional

9 mutually **13** Venn

14 outcome **18** theoretical

20 diagram **21** event

22 die **25** tree

26 dependent **27** step

28 list **29** probability

Down

1 frequency **2** overlapping

4 compound **5** random

10 two way **11** independent

12 relative **15** complementary

16 experimental **17** dice

18 trial **19** expected

23 sample **24** exclusive

Tree diagrams

PAGE 114

1 a All possible arrangements of boys and girls.

b i $\frac{1}{4}$ **ii** $\frac{1}{4}$ **iii** $\frac{1}{2}$

2 a PP, PF, FP, FF **b i** $\frac{1}{4}$ **ii** $\frac{1}{4}$

3 a

Saturday	Sunday	Monday	Outcomes
R	R	R	RRR
		$\bar{R}$	$RR\bar{R}$
	$\bar{R}$	R	$R\bar{R}R$
		$\bar{R}$	$R\bar{R}\bar{R}$
$\bar{R}$	R	R	$\bar{R}RR$
		$\bar{R}$	$\bar{R}R\bar{R}$
	$\bar{R}$	R	$\bar{R}\bar{R}R$
		$\bar{R}$	$\bar{R}\bar{R}\bar{R}$

b $\frac{3}{8}$

4 $\frac{3}{4}$

5 a

		1st Die					
		1	2	3	4	5	6
2nd die	1	0	1	2	3	4	5
	2	1	0	1	2	3	4
	3	2	1	0	1	2	3
	4	3	2	1	0	1	2
	5	4	3	2	1	0	1
	6	5	4	3	2	1	0

b i $\frac{1}{6}$ **ii** $\frac{1}{6}$ **iii** $\frac{1}{3}$ **iv** $\frac{1}{6}$

6 a 1T, 2T, 3T, 4T, 1H, 2H, 3H, 4H

b i $\frac{1}{8}$ **ii** $\frac{1}{4}$

7 $\frac{7}{27}$

8 B

9 a $\frac{1}{3}$ **b** $\frac{2}{3}$ **c** $\frac{1}{2}$

Two-way tables

PAGE 116

1 a 24 **b** 8

c i $\frac{1}{8}$ **ii** $\frac{1}{6}$ **iii** $\frac{2}{3}$

2 a 70 **b** 50 **c i** $\frac{43}{70}$ **ii** $\frac{9}{70}$

d $\frac{9}{25}$

3 a 50 **b** sport

c i $\frac{22}{50} = \frac{11}{25}$ **ii** $\frac{20}{50} = \frac{2}{5}$ **iii** $\frac{1}{10}$

iv $\frac{15}{50} = \frac{3}{10}$

4 a 32 **b** 16

c i $\frac{28}{60} = \frac{7}{15}$ **ii** $\frac{44}{60} = \frac{11}{15}$ **iii** $\frac{7}{60}$

5 a 130 **b** males

c i $\frac{9}{130}$ **ii** $\frac{37}{130}$ **iii** $\frac{116}{130} = \frac{58}{65}$

d It may be too far to walk.

Chapter 11

StartUp assignment 11

PAGE 122

1 a 4.20 **b** \$21 386.70 **c** 678.58 m^3

2 The mode is the most common value.

3 1 : 5000 **4** $b + 3$

5 285° **6** $x = 3\frac{1}{2}$

7 $x = 44$ **8** 10.5 km

9 121.5 cm² **10** 24°

11 17 units **12** \$145.11

13 $\frac{1}{2}$ **14** 133°

15 Example answer given. Other rhombuses are possible. Teacher to check.

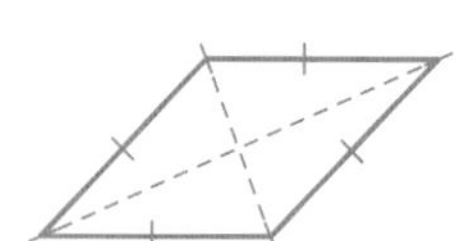

16 $\angle AED$ **17** 360°

18 equilateral

19 a 275 km **b** 6.5 cm

c 1 : 5 000 000

20 Example answer given. Other obtuse-angled triangles are possible. Teacher to check.

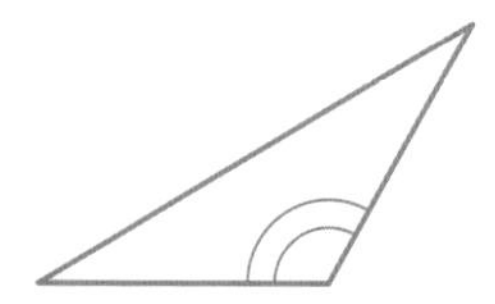

9780170454551

21 a $a = 72$ **b** $y = 110$ **c** $p = 240$
d $e = 60$ **e** $u = 45$ **f** $x = 135$

22 8

23 a false **b** true **c** false

24 25°, 25°

25 a A and B **b** SAS and RHS

26 a $\angle J$ **b** $d = 32\frac{1}{2}$

Challenge:

Other solutions possible.

Proving properties of quadrilaterals PAGE 124

1 a SSS
b Teacher to check matching angles
c axis, bisects
d Teacher to check diagonal
e angles in an isosceles triangle
f AAS
g $\angle AEB$
h 90°
i right angles

2 a AAS
b $\angle YXW$
c Teacher to check matching angles
d equal, equal
e Teacher to check diagonal
f vertically opposite angles
g AAS
h Teacher to check matching sides
i bisect

3 a SSS
b Teacher to check matching angles
c 90°
d right angles, bisect

4 a SAS **b** QS **c** equal

5 a They are equal and bisect each other
b SSS
c base angles of an isosceles triangle
d Teacher to check matching angles
e 45°
f 90°
g equal, right angles, bisect

Geometry crossword PAGE 126

Across

3 ratio
8 nonagon
9 similar
12 quadrilateral
16 bisect
20 matching
25 enlargement
26 SSS
28 exterior
29 symmetry
30 dodecagon
31 interior
32 reduction
33 polygon

Down

1 pentagon
2 rhombus
3 RHS
4 test
5 right
6 AAS
7 diagonal
10 parallelogram
11 trapezium
13 equilateral
14 obtuse
15 decagon
17 triangle
18 eighty
19 isosceles
21 congruent
22 rotational
23 rectangle
24 geometry
26 square
27 SAS or SSS

Geometrical proofs order activity PAGE 132

1 13, 14, 24, 22, 12, 10
2 23, 8, 9, 18, 15, 7, 4, 19
3 11, 20, 6 and 16 (in any order), 5, 17, 21

Chapter 12

StartUp assignment 12 PAGE 133

1 −5
2 $18 000
3 45°
4 2 h 40 min
5 0.000 313
6 5
7 rectangle
8 $\frac{1}{4}$
9 235.62

10

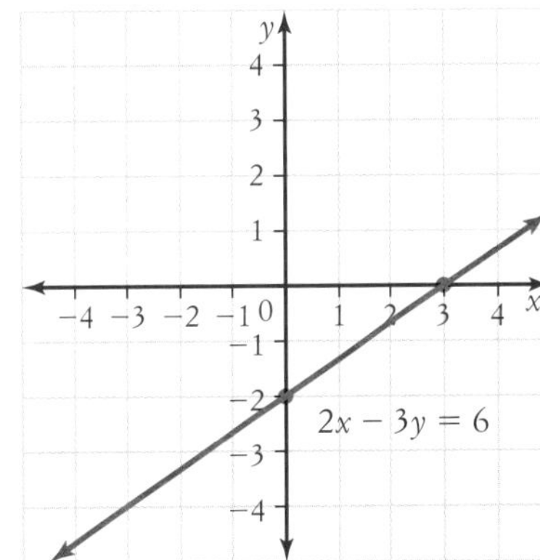

11 SAS

12 $a = \frac{v^2 - u^2}{2s}$

13 $45

14 a 38°13′ **b** 17.3 cm²

15 1, 3, 5, 15

9780170454551

16 $x^2 + y^2 = 16$

17 a parabola **b** -10 **c** -5 and 1

18 a $14m + 12$ **b** $2x^3 - x^2 - 15x$

19 24, remainder 3

20 195

21 a -8 **b** 0 **c** 10

22 a $(3b + 7)(b - 2)$ **b** $2y(y + 3)(y + 4)$

23 a 3 **b** 2 **c** 6

24 25, 5

25 a

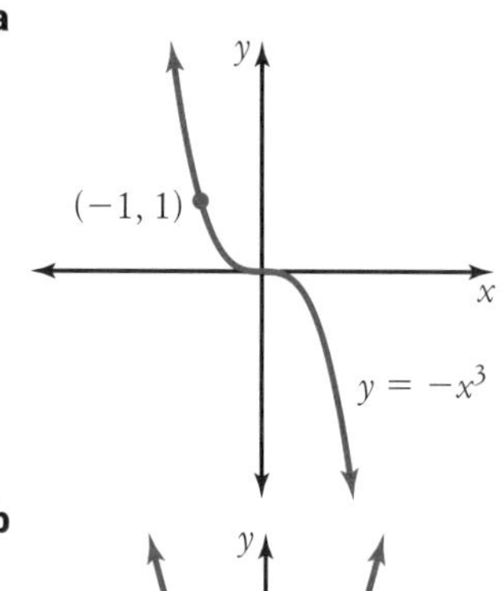

b

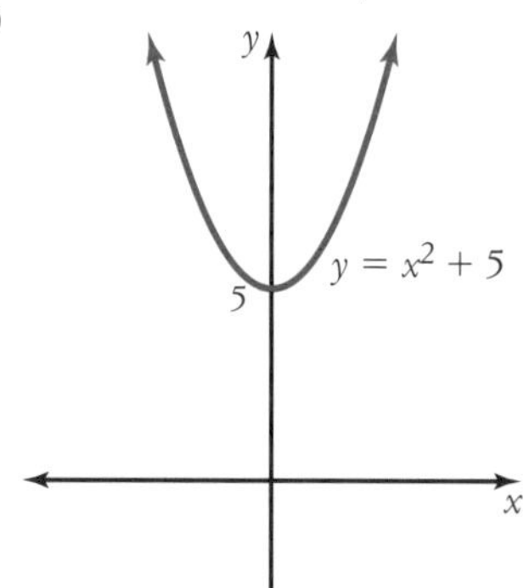

c

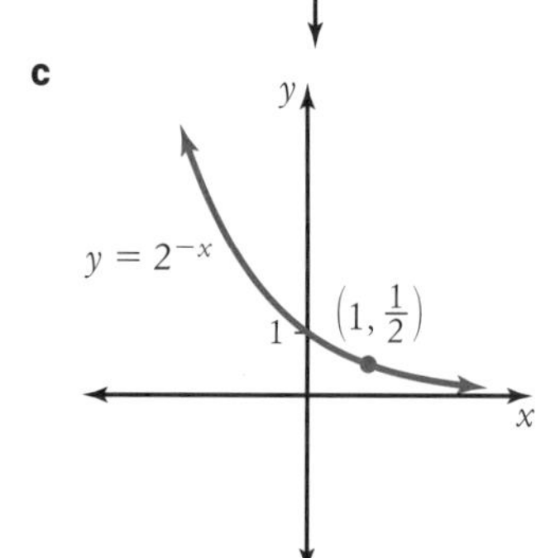

26 a $x = -8$ or 7 **b** $p = 2\frac{2}{3}$

27 $x = -2\frac{2}{5}, y = 6\frac{19}{25}$ or $x = 2, y = 5$

Challenge: 100

Special products PAGE 136

1 $a(a + 4) + 4(a + 4) = a^2 + 4a + 4a + 16 = a^2 + 8a + 16$

2 $p(p + 3) + 3(p + 3) = p^2 + 3p + 3p + 9 = p^2 + 6p + 9$

3 $m(m - 5) - 5(m - 5) = m^2 - 5m - 5m + 25$
$= m^2 - 10m + 25$

4 $k(k - 8) - 8(k - 8) = k^2 - 8k - 8k + 64 = k^2 - 16k + 64$

5 $3p(3p - 7) - 7(3p - 7) = 9p^2 - 21p - 21p + 49$
$= 9p^2 - 42p + 49$

6 $5m(5m + 3y) + 3y(5m + 3y) = 25m^2 + 15my + 15my + 9y^2$
$= 25m^2 + 30my + 9y^2$

7 $f(f + 4) - 4(f + 4) = f^2 + 4f - 4f - 16 = f^2 - 16$

8 $g(g - 3) + 3(g - 3) = g^2 - 3g + 3g - 9 = g^2 - 9$

9 $3p(3p + 7) - 7(3p + 7) = 9p^2 - 21p + 21p - 49$
$= 9p^2 - 49$

10 $9y(9y + 5) - 5(9y + 5) = 81y^2 + 45y - 45y - 25$
$= 81y^2 - 25$

a $w^2 + 12w + 36$ **b** $k^2 - 8k + 16$

c $d^2 - 25$ **d** $36f^2 - 49$

e $4m^2 - 25y^2$ **f** $64t^2 - 9$

g $9m^2 + 6m + 1$ **h** $25m^2 + 70m + 49$

i $100t^2 + 60at + 9a^2$ **j** $4k^2 - 20kx + 25x^2$

k $64 - 80m + 25m^2$ **l** $16 - 24g + 9g^2$

m $9y^2 - 60y + 100$ **n** $16a^2 - 88ab + 121b^2$

Factorising puzzle PAGE 138

1 D	**2** K	**3** R	**4** L	**5** T
6 X	**7** A	**8** Z	**9** Q	**10** I
11 B	**12** S	**13** U	**14** A	**15** M
16 E	**17** G	**18** T	**19** J	**20** Y
21 N	**22** O	**23** P	**24** H	**25** W
26 E	**27** S	**28** C	**29** V	**30** F

What is the difference between a doctor and an algebra student?

The doctor rectifies us while the student factorises.

Chapter 13

StartUp assignment 13 PAGE 142

1 6.5 **2** $3a + 7$

3 a 42.6 m **b** 77.4 m^2

4 62 000 **5** \$625

6 2000 **7** nP

8 $b = 30$ **9** 108°

10 $x = 11$ **11** $\frac{1}{36}$

12 $(3, -2)$ **13** x-axis

14 Circle, centre (0, 0), radius 3 units

15 7 **16** 3.61

17 2 **18** $\sqrt{25}, 0.1\dot{6}, 3.51$

19 $2x^2 + 5x$ **20** $18a^4$

21 8 and 9 **22** 30

23 $3pr - 3p^2$ **24 a** 16, 4 **b** 25, 49

25 $h = \sqrt{13}$ **26** $2x^2 + 7x - 4$

27 $\sqrt{40}, \sqrt{81}, \sqrt{98}, 10.1, 3.6^2$

28 a

29 a 25, 3 **b** 16, 5 or 4, 20

30 No **31** $4k^2 - 25$

32 Yes

33 a $x = \sqrt{3}$ **b** $\frac{\sqrt{3}}{2}$

34 $9x^2 - 6xy + y^2$ **35** Yes

36 $\sqrt{58}$

Challenge: $x = \sqrt{2}$

9780170454551

Surds PAGE 144

1 a $\sqrt{7}, \sqrt{3}$ **b** $\sqrt[3]{100}, \sqrt{10}$
c π, 0.21937... **d** $\sqrt{28}$
2 16, 25, 36, 49, 64, 81, 100
3 a 4, 5 **b** 9, 10
c 6, 7 **d** 7, 8
e 3, 4 **f** 5, 6
4 a 25 **b** 7
c 10 **d** 8
e 12 **f** 4
g 11 **h** 2
5 a 10 **b** 10
c 30 **d** 30
e 18 **f** 18
g 28 **h** 28
i 9 **j** 9
6 a $3\sqrt{5}$ **b** $3\sqrt{2}$
c $4\sqrt{3}$ **d** $10\sqrt{3}$
e $5\sqrt{3}$ **f** $2\sqrt{3}$
g $2\sqrt{2}$ **h** $7\sqrt{2}$
i $2\sqrt{6}$ **j** $4\sqrt{5}$
k $6\sqrt{3}$ **l** $2\sqrt{7}$
7 a $11\sqrt{7}$ **b** $6\sqrt{3}$
c $4\sqrt{5}$ **d** $\sqrt{11}$
e $9\sqrt{6} - \sqrt{10}$ **f** $6\sqrt{2} - 2\sqrt{3}$
g $5\sqrt{2}$ **h** $8\sqrt{3}$
i $2\sqrt{5}$ **j** $2\sqrt{3}$

Simplifying surds PAGE 146

3.141 592 653 589 793 238 462 643

Rationalising the denominator PAGE 147

1 $\frac{5\sqrt{2}}{2}$ **2** $\frac{\sqrt{6}}{2}$ **3** $\frac{\sqrt{15}}{3}$
4 $\frac{\sqrt{2}}{2}$ **5** $\sqrt{6}$ **6** $\frac{5\sqrt{30}}{12}$
7 $\frac{2\sqrt{5}+\sqrt{15}}{5}$ **8** $\frac{\sqrt{30}-3\sqrt{6}}{6}$ **9** $2\sqrt{3}+2$
10 $\frac{7\sqrt{5}-5}{15}$ **11** $2\sqrt{10}-\sqrt{6}$ **12** $\sqrt{70}+3$
13 $\frac{4\sqrt{2}-\sqrt{3}}{2}$ **14** $\frac{7\sqrt{10}+15\sqrt{2}}{30}$
15 $\frac{30-6\sqrt{2}}{23}$ **16** $\sqrt{5}+1$
17 $\frac{3\sqrt{2}-2\sqrt{3}}{2}$ **18** $\frac{5\sqrt{6}+\sqrt{30}}{5}$
19 $2\sqrt{6}+3\sqrt{2}$ **20** $\frac{13+5\sqrt{2}}{7}$

Surds crossword PAGE 152

Across

5 difference **8** square
9 approximate **13** rationalise
15 root **16** real

Down

1 denominator **2** expression
3 binomial **4** undefined
6 irrational **7** quotient
8 simplify **10** product
11 expand **12** rational
14 surd

Chapter 14

StartUp assignment 14 PAGE 153

1 0.707 **2** $\frac{1}{8}$
3 10 000 000 **4** 198
5 Teacher to check. The equation must be of the form $y = mx + c$ where $m < -1$ and $c > 0$
6 $5y$ **7** 86.32 cm^2
8 6150 **9** rectangle
10 108 **11** $\frac{3}{10}$
12 34° **13** \$1300, \$650, \$2600
14 360° **15** \$674.66
16 a 3 **b** -2
17 a 14 **b** 22 **c** -5 **d** 0
18

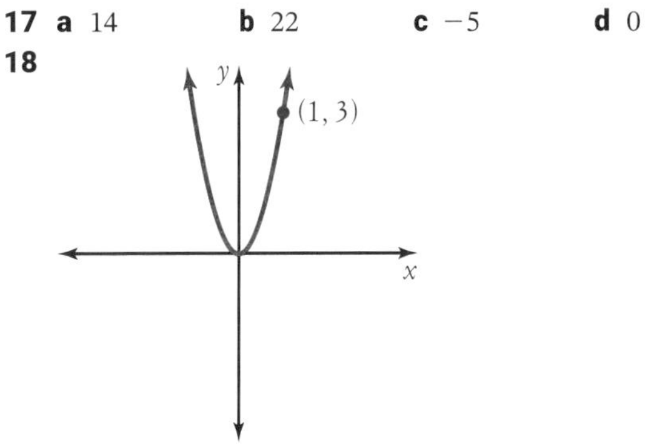

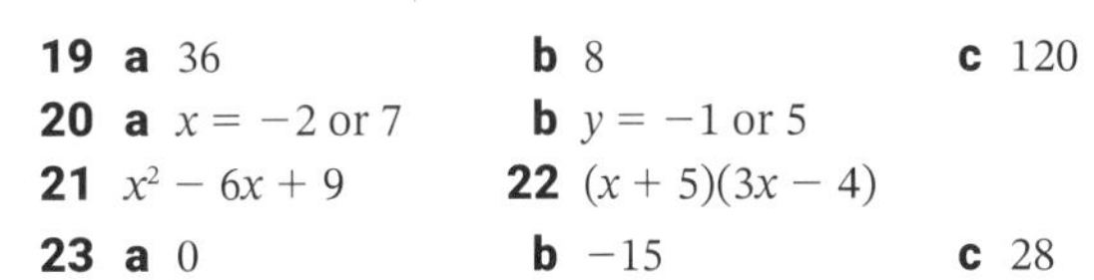

19 a 36 **b** 8 **c** 120
20 a $x = -2$ or 7 **b** $y = -1$ or 5
21 $x^2 - 6x + 9$ **22** $(x + 5)(3x - 4)$
23 a 0 **b** -15 **c** 28
24

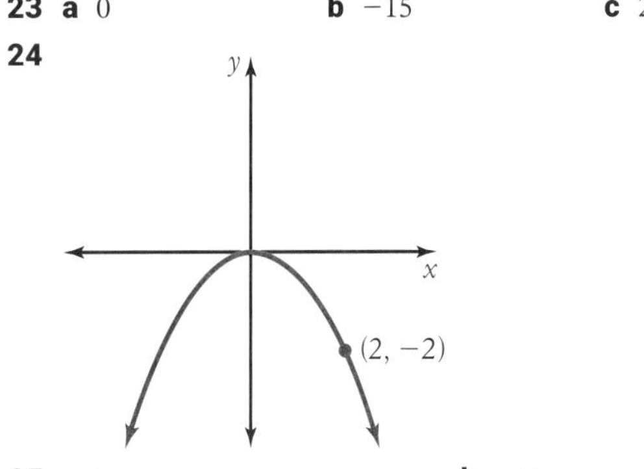

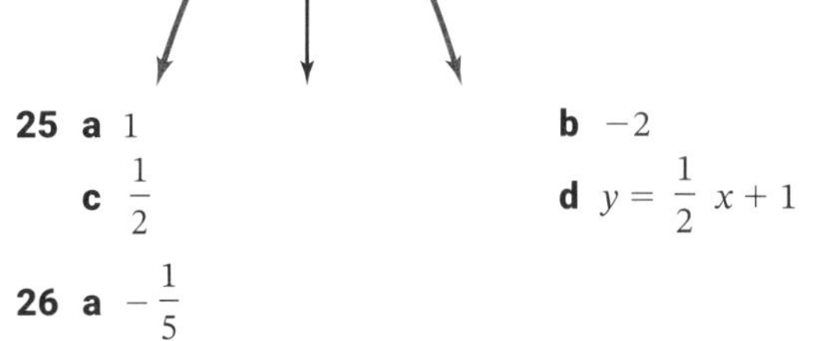

25 a 1 **b** -2
c $\frac{1}{2}$ **d** $y = \frac{1}{2}x + 1$
26 a $-\frac{1}{5}$
27 a 2.83 **b** 5
Challenge: 2

Quadratic equations puzzle PAGE 156

1 W **2** G **3** Q
4 X **5** O **6** C
7 E **8** V **9** H
10 D **11** M **12** P
13 A **14** F **15** U
16 T **17** J **18** I
19 S **20** B **21** O
22 E **23** A **24** T
25 I **26** Y **27** K
28 R **29** L **30** N

Playing this game is like graphing lines and curves on a number plane. You've got to look for the intercepts.

Quadratic equations crossword PAGE 158

Across

2 simultaneous
4 factorise
5 coordinates
9 constant
11 concave
14 symmetry
16 formula
17 parabola
19 intercept
21 solution
23 intersection

Down

1 coefficient
3 quadratic
6 monic
7 root
8 square
9 complete
10 vertex
12 equation
13 hyperbola
14 substitution
15 surd
18 axis
20 exact
22 circle

Chapter 15

StartUp assignment 15 PAGE 164

1 \$250
2 8
3 1 h 40 min
4 $\frac{8}{11}$
5 48 m
6 $14 + 6\sqrt{5}$
7 201.84cm^2
8 \$9889.05
9 $-\frac{3}{4}$
10 $(4y + 3)(y - 5)$
11 $d = 32$
12 $k \leq \frac{1}{3}$
13 4.5
14 $y = -3$
15 $y = -\frac{1}{3}x$
16 $\frac{a}{b}$
17 **a** 1.1276 **b** 5.7671 **c** 1
d 0.3420 **e** 0.3420
18 **a** same **b** 40°
19 **a** $\frac{\sqrt{3}}{2}$ **b** $\frac{1}{\sqrt{3}}$
20 **a** 30°58′ **b** 56°26′ **c** 60°
d 61°56′
21

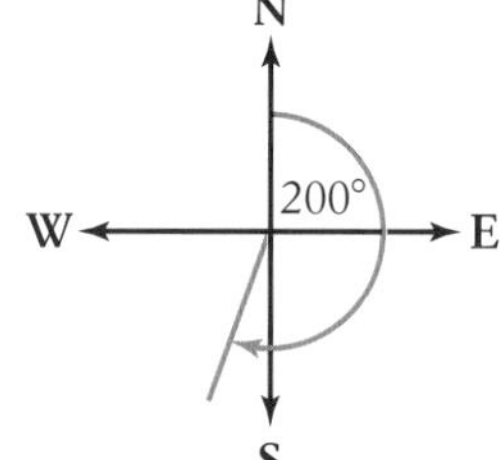

22 **a** 15.43 **b** 26.46
23 **a** 86.24 **b** 6.47 **c** 15.17
d 2.30
24 502.51 m
25 20°
26 **a** 10 km **b** 323°8′
Challenge: $\theta = 45°$

Finding an unknown angle PAGE 167

1 33° **2** 68°
3 58° **4** 80°
5 115° **6** 124°
7 26° **8** 66°
9 17° **10** 27°
11 136° **12** 137°
13 83° **14** 63°
15 125° **16** 117°
17 90° **18** 90°
19 60° **20** 70°
21 88°

CHAPTER 1

Compound interest

PAGE 9

Part A

1 \$200
2 8 and 9
3 $\sqrt{82}$
4 a $SA = 2\pi r^2 + 2\pi rh$
b $V = \pi r^2 h$
5 a 4
b 3
c 3

Part B

1 a 0.175
b 0.05
2 \$4612.50
3 a 52
b 12
c 365
4 a \$63 814.08
b \$68 106.54

Part C

1 \$552.08
2 6.5%
3 \$20 058.66
4 a \$23 936.29
b \$3 936.29
5 \$581.46

Part D

1 a simple interest
b compound interest
c the number of periods
d principal
2 a \$3325
b \$41 325
3 a \$16 766.12
b \$266.12

CHAPTER 2

Coordinate geometry

PAGE 18

Part A

1 \$500
2 15%
3 a 1000
b 0.24 or $\frac{6}{25}$
4 $n = 65$ (alternate angles on parallel lines), $m = 25$ (angles in a right angle)

Part B

1 a

x	−2	−1	0	1
y	−4	−3	−2	−1

b

x	−1	0	1	2
y	−7	−3	1	5

2 a neither
b positive
c negative
d neither

Part C

1 a $\sqrt{20}$ or $2\sqrt{5}$
b (4, 2)
c $\frac{1}{2}$

2 a

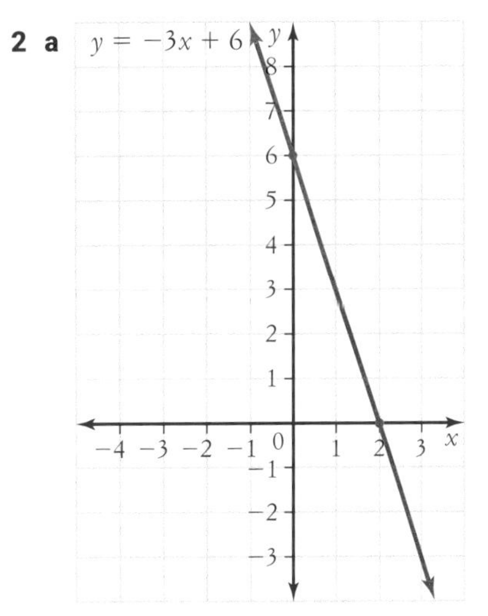

x-intercept 2, y-intercept 6.

3 −3
4 It does.

Part D

1 a

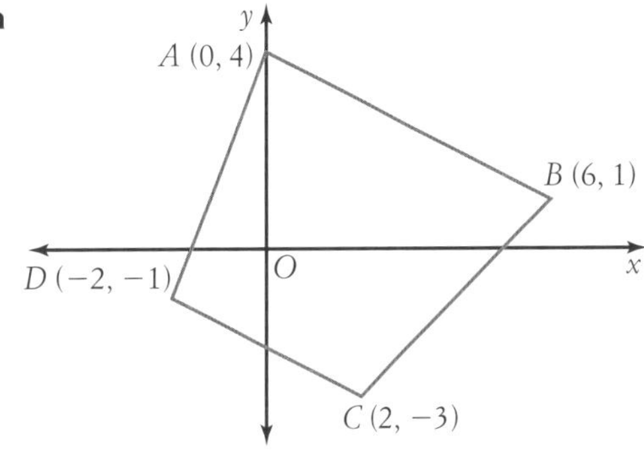

b $m_{AB} = -\frac{1}{2}, m_{CD} = -\frac{1}{2}, m_{AD} = \frac{5}{2}, m_{BC} = 1$
c trapezium
2 $y = -x - 2$
3 $x = 3$

Graphing lines

PAGE 20

Part A

1 a $7m(7m - 5n)$
b $(x + 2)(8 - p)$
2 a $14x^4y^2$
b $\frac{16m^4}{9}$
3 a 69
b 30s and 40s
4 $W = 21$

Part B

1 a 2
b $-\frac{1}{2}$
2 $y = \frac{1}{2}x + 7$
3

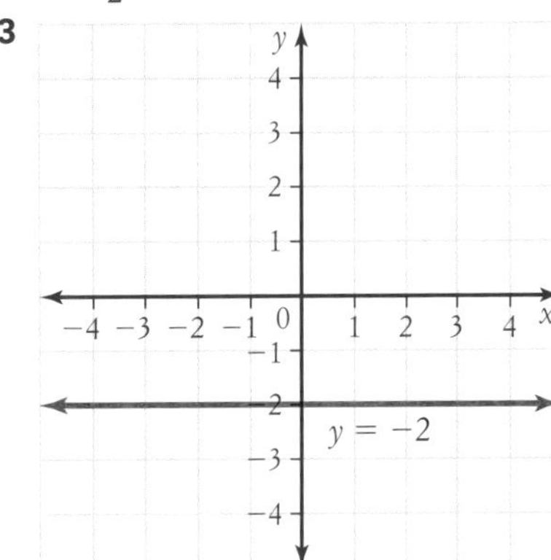

4 **a** $m_1 = 4, m_2 = 4$: parallel because $m_1 = m_2$

b $m_1 = -\frac{1}{2}, m_2 = 2$: perpendicular because $m_1 = \frac{-1}{m_2}$

Part C

1 $y = -3x - 3$

2 **a** $y = 5x + 3$ **b** $y = x - 1$

3 **a** $y = -\frac{1}{5}x + 3$ **b** $y = \frac{1}{3}x - 7$

Part D

1 **a** parallel **b** perpendicular

2 **a** $\frac{3}{2}$ **b** $(1, -1)$

c $m = -\frac{2}{3}$ **d** $y = -\frac{2}{3}x - \frac{1}{3}$ or $y = \frac{-2x-1}{3}$

3 **a** $y = \frac{1}{2}x + 2$ **b** $y = -2x - 3$

CHAPTER 3

Surface area

PAGE 33

Part A

1 $x = 7$

2 $37.50

3 **a** $\frac{56}{65}$ **b** $\frac{56}{65}$ **c** $\frac{33}{56}$

4 **a** Monday **b** Wednesday

5 2.5 cm

Part B

1 **a** 384 units^2 **b** 336 units^2

c 706.86 units^2 **d** 16 cm^2

e 326.73 cm^2

2 **a** $x = 12$ **b** $y = 3.9$

Part C

1 **a** 396 cm^2 **b** 792 cm^2

2 **a** 151.33 m^2 **b** 251.33 m^2

Part D

1 The total area of all the faces of the solid.

2 cross-section, polygon

3 $2\pi rh$

4 **a** 1894.5 m^2 **b** 14.4 m^2

Volume

PAGE 34

Part A

1 $x^2 + 15x + 36$

2 $186

3 28.8 km/h

4 **a** 1.5 **b** 7.5 cm

5 $y = 2x - 5$

6 14

Part B

1 310 cm^2

2 257.59 cm^2

3 240.27 cm^2

Part C

1 **a** 12 cm^3 **b** 6283 cm^3

c 504 cm^3 **d** 128 mm^3

2 **a** 560 m^3 **b** 909.43 mm^3

Part D

1 **a** volume **b** cubic

c capacity **d** cubic centimetre

2 **a** 219.1 mm^3 **b** 320 cm^3

CHAPTER 4

Index laws

PAGE 42

Part A

1 $2 \times 2 \times 2 \times 2$, 16

2 0.085

3 8×10^{-4}

4 $\sqrt{73}$

5 $P(\text{K or Q}) = \frac{2}{13}$

6 8, 8

Part B

1 **a** x^3y^2 **b** $25p^3t^2$

2 **a** 729 **b** 100 000

c 64 **d** 1

3 **a** 3 **b** 4

Part C

a $14x^5y^{10}$ **b** $\frac{p^{10}}{81y^2}$

c -2 **d** $64n^6$

e $\frac{1}{x^5}$ **f** $\frac{1}{1331p^3}$

g $10p$ **h** $\frac{25a^2}{4}$

Part D

1 **a** power, or exponent

b The number of times a base is multiplied by itself.

2 **a** a^{m-n} **b** a^nb^n

c $\frac{1}{a}$

3 **a** $\frac{2}{3}$ **b** 1

4 $8p^{-3}$

Algebraic fractions

PAGE 44

Part A

1 $21a^3$

2 $100x^2 - 4y^2$

3 $y = -2x + 3$

4 **a** $\frac{1}{10}$ **b** $\frac{1}{2}$ **c** $\frac{2}{5}$

5 $\frac{3}{4}$

Part B

1 **a** $\frac{5}{21}$ **b** $\frac{23}{18}$ **c** $\frac{15}{28}$ **d** $\frac{2}{3}$

2 **a** $\frac{mn^3}{6}$ **b** $\frac{2}{t}$ **c** $25x^6y^4$ **d** $81a^4b^8$

Part C

1 $\frac{17x}{30}$

2 $\frac{y}{48}$

3 $\frac{2m}{35}$

4 $\frac{13y}{20}$

9780170454551

5 $\frac{12y^2}{x^2}$ 6 $\frac{m^2}{12}$ 7 $\frac{3p}{35a}$ 8 $\frac{4}{15n}$

Part D

1 a denominator, numerators
b numerators, denominators
c reciprocal
2 6
3 a 3^{-2} is the same as $\frac{1}{3^2}$, so the final answer is $\frac{1}{9}$
b Any number to the power of 0 is 1, so $3^0 = 1$

Expanding and factorising
PAGE 46

Part A

1 a $5y^2 - 9y - 6$ b $-27x^2 + 7x^3$
2 $A = \pi r^2$
3 $2y(5 - 2y)$
4 0.811
5 27
6 22%
7 25

Part B

1 a $3x^4$ b $30a^9$
c $\frac{m}{9n}$ d $\frac{7}{6}$
2 $-32y^2 - 104y$
3 a 4, 9 b -8, 9
4 $-8(3b - 2)$

Part C

1 a $20xy^2 - 32x^2y$ b $-18a - 6a^2$
c $10p^3 + p^2$ d $23ab + 30a - 54b$
2 a $12x(3 + 2x)$ b $-2y(4x - y)$
c $(p - q)(3 + p)$ d $8a^2b(2a + 1)$

Part D

1 a $xy + xz$ b $kr - ks$
2 a highest common factor b $2n$
c $2n(4m - n)$
3 divide
4 $-2b^2 - 2ab$

Binomial products
PAGE 48

Part A

1 $x = \pm 4$
2 $S = 10$
3 $x = 65$
4 a 70 m b 146 m^2
5 a $\frac{1}{12}$ b $\frac{5}{12}$

Part B

1 a $6x^2y^2 - 6xy$ b $-16a^3 - 12ab$
c $-3m^2 - 10m + 8$ d $16a^2 - 52a + 42$
2 a $4xy(2y + 1 + 11x)$ b $-a(3a^4 + 6a^3 - 1)$

Part C

1 a $x^2 + x - 12$ b $2y^2 + 2y - 60$
c $2n^2 + 3n - 35$
2 a $(x - 7)(x + 9)$ b $(y + 4)(y - 8)$

Part D

1 An algebraic expression with terms, $x + 5$
2 -27
3 An algebraic expression where the highest power is 2, $x^2 - 9x + 14$
4 -7 and 2
5 a $5x(5x + y)$ b $-9p(9 - p)$
c $3ab(b - 12a)$ d $(2x - 5)(3x - 5)$

CHAPTER 5

Data
PAGE 58

Part A

1 $\frac{7}{12}$
2 a $-\frac{n}{10}$ b $\frac{7}{9}$
3 475 km
4 a $5\sqrt{44}$ or $10\sqrt{11}$ cm^2 b $\frac{26\pi}{9}$ m^2

Part B

1 a 4 b 3.06
c 3 d 3, 5
2 a 34 b 34.9
c 27 d 53

Part C

1 a symmetrical b positively skewed
c negatively skewed
2 a 13 b 17 c 22.5
d 9.5 e 16, 17

Part D

1 a spread b left, lower values
c bimodal d outlier
e Q_1, Q_U or lower quartile
2 a negatively skewed b 8
c $9.5 - 6 = 3.5$

Boxplots
PAGE 60

Part A

1 a 2 b -1
2 $\frac{3w^2}{25z}$
3 a 5 b $\frac{4}{5}$
4 (other reasons possible)
$\angle FEG = \angle CBE = 130°$ (corresponding angles, $AC \parallel DF$)
$x = 180 - 130 = 50$ (angles on a straight line)

Part B

1 48
2 36
3 36.11
4 10
5 26
6 49
7 58
8 23

Part C

1 a 0 b 1.5
c 3 d 5.5
e 9 f 4
2

Part D

1 25%, 50%, 25%

2 **a** highest value **b** lower quartile, Q_L or Q_1

c median

3 Lowest value, lower quartile, median, upper quartile, highest value

Comparing data PAGE 62

Part A

1 **a** 8 **b** $\frac{15}{17}$ **c** $1\frac{3}{5}$

2 $200

3 **a** $-27a^3b^6$ **b** $\frac{3}{a}$

4 $x = 6\frac{1}{2}$

Part B

1 **a** 20.5 **b** 13

2

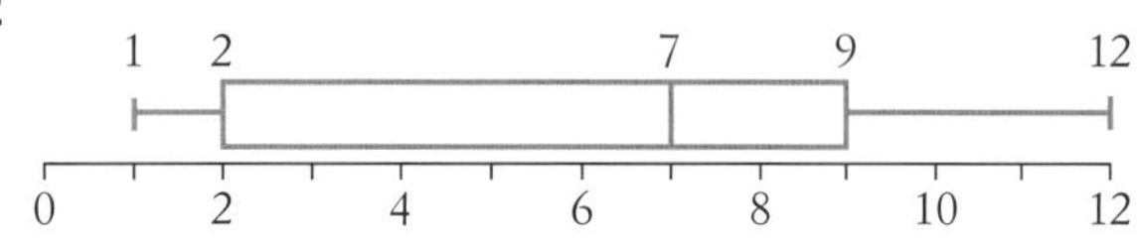

Part C

1 **a**

	Median	Interquartile range
Boys	69	4
Girls	65.5	5

b Boys are generally taller than girls.

2 **a** 22.875 **b** 22.75

c Ages of females are more spread out, including an outlier of 103.

Part D

1

Age group	Ashfield Frequency	Burwood Frequency
28– < 33	1	0
33– < 38	2	1
38– < 43	2	0
43– < 48	6	4
48– < 53	6	5
53– < 58	2	6
58– < 63	1	3
63– < 68	0	1

2 Ashfield: symmetrical, Burwood: negatively skewed

3 Burwood's teachers are generally older

Scatterplots PAGE 64

Part A

1 46.09, 46.199, 46.2

2 6 : 1

3 **a** $120 000 **b** 60%

4 **a** $x = -1$ **b** $\frac{1}{22}$

Part B

1 Data 2 Geometry

3 Geometry 4 Algebra

5 Data 6 Geometry

7 Data 8 Algebra

Part C

1 **a** strong negative **b** weak positive

c no relationship

2 **a**

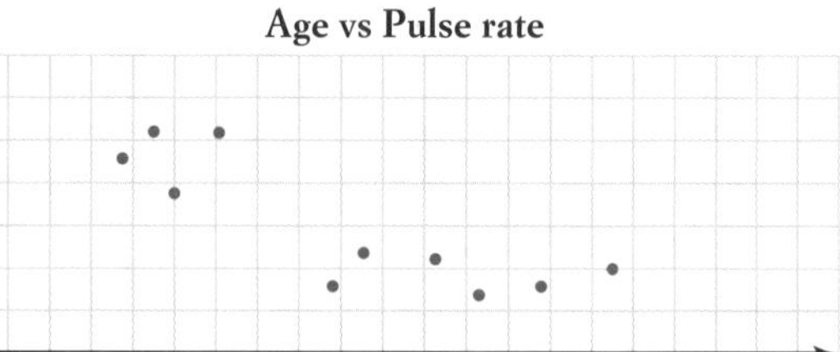

b negative **c** decreases

Part D

1 dependent 2 bivariate data

3 positive

4 **a** strong **b** perfect

5 **a** negatively skewed **b** 8

c 34

CHAPTER 6

Equations PAGE 72

Part A

1 **a** $\frac{3m}{5}$ **b** 12

c $\frac{56}{q}$ **d** $\frac{9}{t^5}$

2 **a** $147 **b** $86.40

4 $x = 55$, co-interior angles on parallel lines

Part B

1 **a** $y = 2\frac{1}{2}$ **b** $m = 63$ **c** $b = 26$

2 54

Part C

1 $x = -\frac{1}{6}$ 2 $m = 7\frac{1}{2}$

3 $a = 11$ or 7 4 $y = -4$ or 5

Part D

1 2 2 Finding its square root.

3 **a** There are 2 numbers, 2 and −2, that when squared gives 4.

b There isn't a number that can be squared to give a negative answer. You cannot take the square root of a negative number.

4 **a** $k = 0$ or -4 **b** $y = 11\frac{1}{2}$

Equations and formulas PAGE 74

Part A

1 **a** $64 **b** $864

2 **a** $-5n(4m - 5n)$ **b** $-(x + 4)(3 + y)$

9780170454551

3 a 115°, supplementary angles add to 180°
b 65°, vertically opposite angles are equal.

Part B

1 a $a = 6\frac{3}{7}$ **b** $x = -3\frac{1}{3}$

2 a $x = 5$ cm **b** 101 cm^2

Part C

1 77°F

2 a $x = 17$ **b** 83°

3 a 302.5 cm^2 **b** 14.9 cm

Part D

1 5 **2** 16

3 20

Inequalities

PAGE 76

Part A

1 15 **2** 130

3 $0.\dot{1}2\dot{4}$

4 a 22, 24 **b** 24.5

5 a 84 m^2 **b** 300 mm^2

6 Diagonals bisect each other.

Part B

1 a x is greater than 4; for example, 9
b x is less than or equal to 4; for example, 0

2 a −3 −2 −1 0 1 2 3 x

b 0 1 2 3 4 5 6 7 8 9 10 11 x

c −4 −3 −2 −1 0 1 2 3 4 x

d −1 0 1 2 3 4 5 x

Part C

1 a $x > 7$ **b** $x \le -4$

2 a $m \le 7$ **b** $x > -1$
c $y \ge -34$

Part D

1 solution **2** true

3 multiplying (or dividing), reverse

4 a $m \le -18$ −18 m

b $x < -\frac{1}{2}$

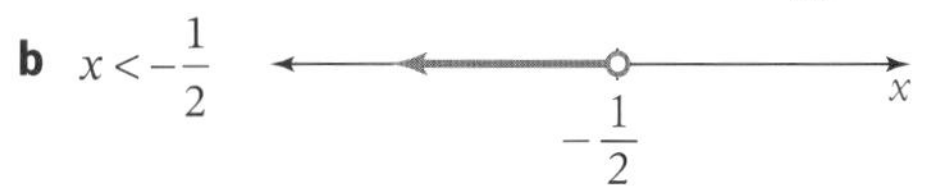

CHAPTER 7

Graphing curves

PAGE 86

Part A

1 62.5%

2 a $x^2 - x - 12$ **b** $\frac{a^2b}{3}$ **c** $\frac{1}{36m^2}$

3 square, rhombus

4 a 80% **b** 5

Part B

1 a $R = 0.74N$ **b** 3.7 **c** Straight line, 0

2 980.87

3 a $y = 8$ **b** $y = 1$
c $y = 4, -4$

Part C

1 a C **b** A
c B

2 a D **b** C
c B **d** A

3 $x^2 + y^2 = 16$

Part D

1 a Concave down, the coefficient of x^2 is negative ($a = -3$)
b 1

2 centre (0, 0), radius 3

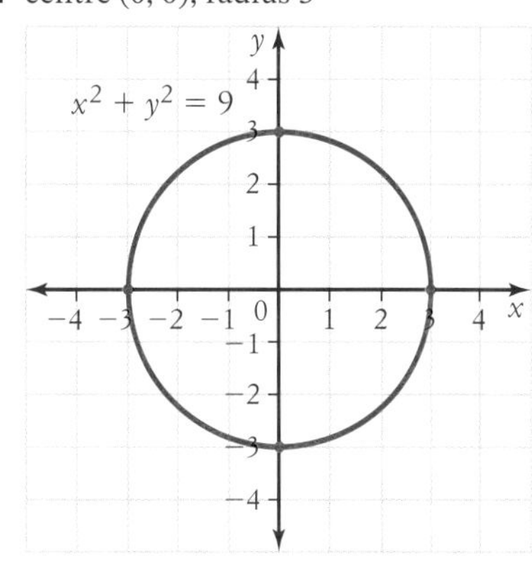

3

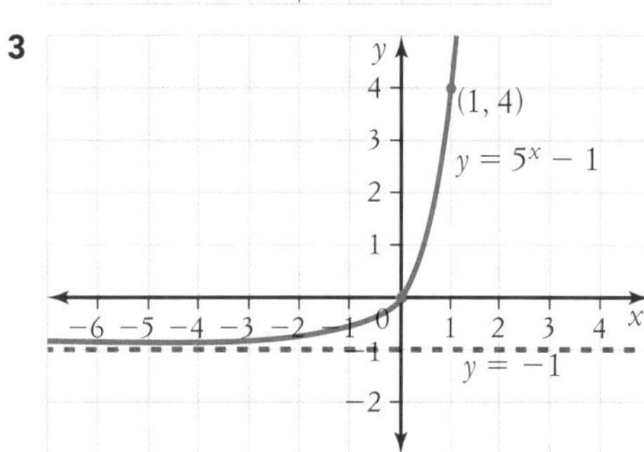

Asymptote $y = -1$.

4 Vertex (0, 1)

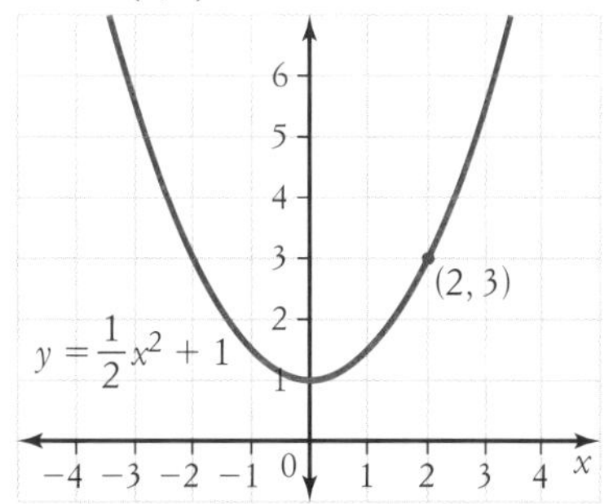

CHAPTER 8

Trigonometry 1

PAGE 96

Part A

1 a 0.82 **b** 0.075

2 $d = 6$

3 $x = 80, y = 130$

4 a 1 **b** $\frac{9p^2}{4x^2}$

5 −4°F

Part B

1 a $x = 27.66$ b $y = 33.99$

2 a $\theta = 49°$ b $\theta = 48°$

3 a Right-angled if $17^2 = 8^2 + 15^2$.
$289 = 64 + 225$. ∴ It is right-angled.
b Right-angled if $25^2 = 6^2 + 23^2$.
$625 \neq 36 + 529$. ∴ It is not right-angled.

Part C

1 22.69

2 80° 9′

3 a $r = 14.49$ b $h = 25.13$ c $x = 98.88$

4 $\frac{77}{36}\left(\text{or } 2\frac{5}{36}\right)$

5 322 mm

Part D

1 a The longest side of a right-angled triangle
b The trigonometric ratio $\frac{\text{opposite}}{\text{adjacent}}$ for an angle in a right-angled triangle
c $\frac{1}{60}$ of a degree for angle size measurement

2 $A = 46° 24'$, $C = 43° 36'$

3 a $AC = 46.4$ cm b $BC = 42.8$ cm

Trigonometry 2 PAGE 98

Part A

1 a $a = 1\frac{9}{13}$ b $x = 3\frac{21}{43}$

2 \$144

3 $x = 6$, $y = 32$

4 2.07×10^{-5}

Part B

1 a $q = 11.61$ b $b = 16.5$

2 11.7 m

3 a 67° b 56°

4 30° 38′

Part C

1 11.7 m

2 a 253° b 65° c 312°

3 a 5.6 km b 020°

Part D

1 a down, horizontal b Three, true, clockwise

2 225°

3 19°

CHAPTER 9

Simultaneous equations 1 PAGE 106

Part A

1 15

2 a numerical, discrete b numerical, continuous c categorical

3 a $6a^2$ b $27p^2$

4 16

5 $8x^6y^9$

Part B

1 a -7 b 2.5

2

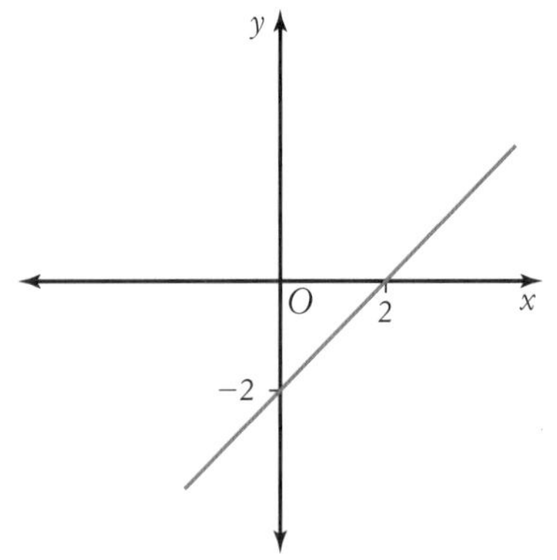

x	−1	0	1	2
y	−3	−2	−1	0

3 a no b yes

4 a Gradient $\frac{3}{5}$, y-intercept -3

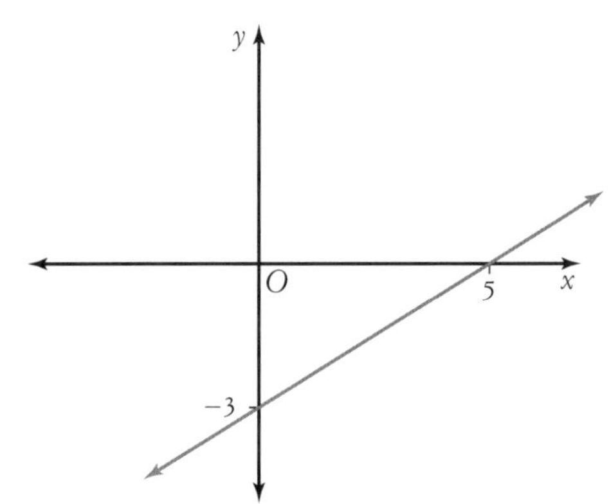

b Gradient $-\frac{1}{2}$, y-intercept 3

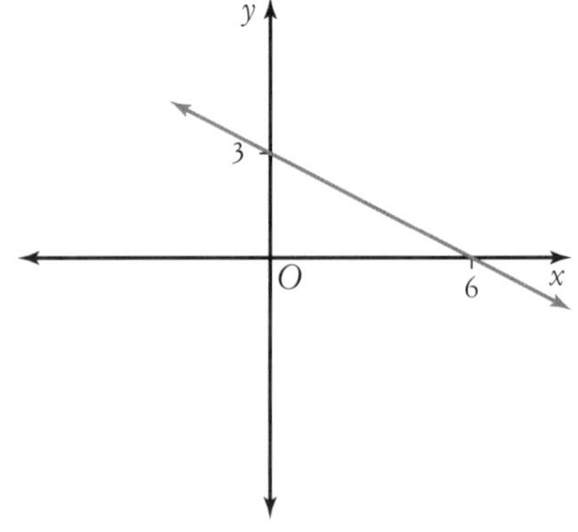

Part C

1 $2x + y = 3$

x	0	1	2	3	4
y	3	1	−1	−3	−5

$x + y = -1$

x	0	1	2	3	4
y	−1	−2	−3	−4	−5

 9780170454551

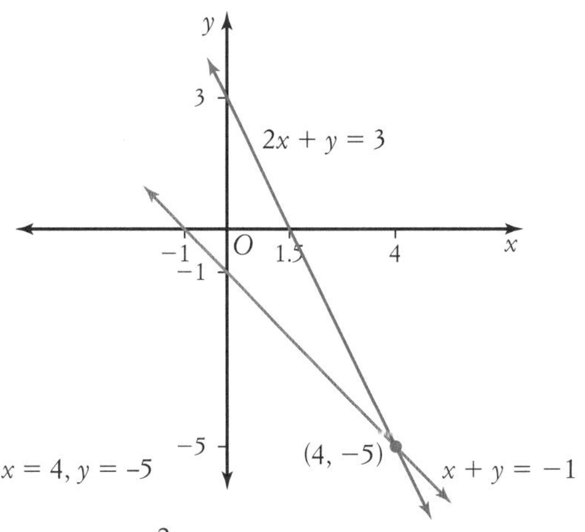

2 $m = 2, n = 7\frac{2}{5}$

Part D

1 Elimination, substitution

2 The coordinates of their point of intersection gives the solution.

3 Substitute the solution back into the simultaneous equations to see if they are correct.

4 $x = 3, y = -1$

Simultaneous equations 2

PAGE 109

Part A

1 103° 46′

2 $-8x^6$

3 $2a^2 + 13a + 15$

4 132 m²

5 8 h 5 min

6 **a** 1 : 400 **b** 18 : 1

7 $4(y - 2)(y - 4)$

Part B

1 $a = 7, b = 9$

2 $m = 5, n = -2$

Part C

1 **a** $x = -1, y = 7$

2 Adult: $80, Child: $60

Part D

1 substitution

2 E, D, C, B, A

3 $x = 10, y = 4$

CHAPTER 10

Probability 1

PAGE 118

Part A

1 $x = 75$ (vertically opposite angles), $y = 105$ (angles on a straight line)

2 $x = -22$

3 centre (0, 0), radius 4

Part B

1 0.25

2 **a** 3

b No: $P(\text{red}) = \frac{1}{5}$, $P(\text{yellow}) = \frac{3}{10}$, $P(\text{blue}) = \frac{1}{2}$

3

	Dog	Cat	Others	Total
Girls	10	18	7	35
Boys	15	10	9	34
Total	25	28	16	69

4 5%

5 100

Part C

1 $\frac{8}{25}, \frac{2}{25}$

2 **a** 9 **b** 54

3

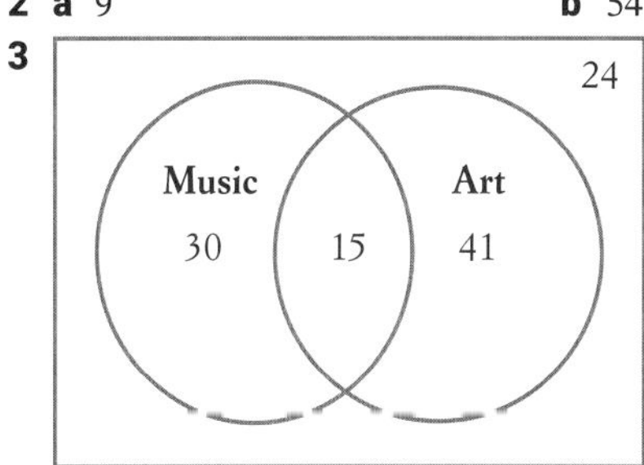

Part D

1 outcomes, outcomes

2 event

3 expected frequency

4 **a** 150 **b** **i** $\frac{7}{25}$ **ii** $\frac{11}{150}$ **iii** $\frac{6}{15}$

Probability 2

PAGE 120

Part A

1 63°43′

2 $\frac{a^{16}b^{30}}{3}$

3 **a** $\frac{6}{5}\left(\text{or } 1\frac{1}{5}\right)$ **b** 12.5

4 $(36 - 9\pi)$ cm²

5 $y = -1\frac{7}{10}$

6 2.5365×10^7

Part B

1 135

2 **a** 31 **b** 4

c

	Cake	Not cake
Lollies	9	22
Not lollies	15	4
Total	24	26

3 **a** $\frac{9}{40}$ **b** $\frac{1}{5}$

Part C

1 **a**

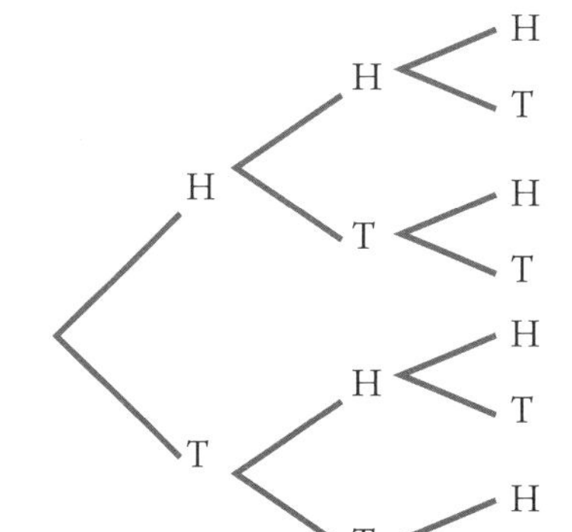

b 8 **c** $\frac{7}{8}$

2 a $\frac{3}{5}$ **b** $\frac{1}{3}$

3 $\frac{1}{6}$

Part D

1 Venn diagram **2** 36

3

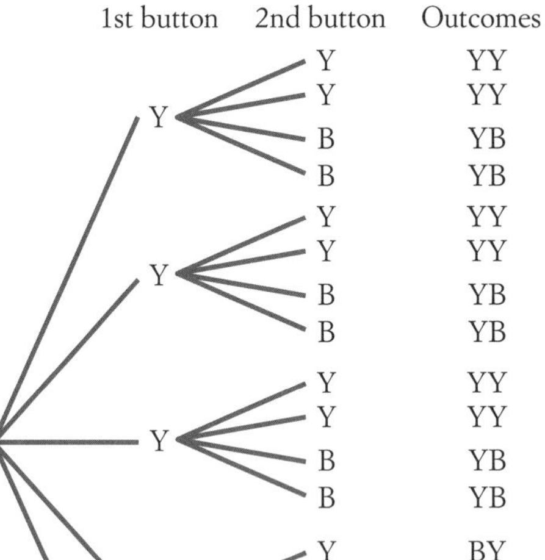

a $\frac{3}{10}$ **b** $\frac{2}{5}$

c $\frac{1}{2}$ **d** $\frac{1}{4}$

CHAPTER 11

Congruent figures

PAGE 128

Part A

1 $\sqrt{19}$ **2** \$144

3 6 : 8 : 3 **4** $(x + 7)(x - 3)$

5 \$2100 **6** 120°

Part B

1 a RHS **b** RP

c $\angle R$ **d** RPM

2 a SSS **b** SAS

3 $d = 42, y = 28$

Part C

1 In $\triangle ABX$ and $\triangle ACX$

$AB = AC$ (given)

$BX = XC$ (given)

AX is common

$\therefore \triangle ABX \equiv \triangle ACX$ (SSS)

2 AXC, 90 **3** $\angle C$

4 2 equal angles (opposite the equal sides)

Part D

1 In $\triangle DAB$ and $\triangle CBA$

$DA = CB$ (opposite sides of a rectangle)

$\angle DAB = \angle CBA = 90°$ (angles of a rectangle)

AB is common

$\therefore \triangle DAB \equiv \triangle CBA$ (SAS)

2 $\angle BCA$ **3** CA

4 Diagonals are equal

5 included angle

Similar figures

PAGE 130

Part A

1 a 1 : 6 **b** 14 : 1

2 $2\frac{6}{25}$ **3** $-6ab^3$

4 $x = 35, y = 55$

5 a 36 and 38 **b** 38

Part B

1 a RHS **b** $\angle CAD$

c In $\triangle ABD$ and $\triangle ACD$

$AB = AC$ (given)

$\angle ADB = \angle ADC = 90°$ ($AD \perp BC$)

AD is common

$\therefore \triangle ABD \equiv \triangle ACD$ (RHS)

2 a RHS **b** $\frac{3}{4}$

Part C

1 a $a = 7.5, b = 20$ **b** $y = 8.75, p = 6.4$

c $c = 16, d = 21$

2 triangle, proportional, triangle

Part D

1 a 1 **b** ratio, equal

2 a corresponding angles on parallel lines, $DE \parallel BC$

b ADE

c AA (equiangular) **d** $x = 9$

CHAPTER 12

Special binomial products

PAGE 140

Part A

1 −27 **2** Pythagorean triad

3 −4 **4** $0.1\dot{6}$

5 155° **6** $14a + 10b$

7 $m = 68, y = 112$

Part B

1 a $p^2 + 3p - 4$ **b** $y^2 - 12y + 27$

2 a $(x - 3)(x + 11)$ **b** $(a - 4)(a - 7)$

Part C

1 a $x^2 - 16x + 64$ **b** $4k^2 + 20k + 25$

2 a $(2a - 3c)(b + d)$ **b** $(3d + 2)(2d - 1)$

c $(2y + 9)(2y - 9)$

Part D

1 Teacher to check, for example:

a $(x + 1)^2$ **b** $(x + 1)(x - 1)$

2 a $9y^2 - 49$ **b** $9y^2 - 42y + 49$

3 $3(c + 2)(c + 6)$

CHAPTER 13

Surds 1 PAGE 148

Part A

1 a 24, 32 **b** 30

2 $m = 5\frac{3}{4}$ **3** $x = 110$

4 (0, 0), 6 units **5** 256 cm^2

6 (1, 7) **7** discrete

Part B

1 49, 100

2 a $-18p - 24$ **b** $x^2 - 10x + 21$

3 a x **b** 5

4 $\sqrt{88}$ $\sqrt{125}$

5 a irrational **b** rational

Part C

1 a $4\sqrt{10}$ **b** $\frac{\sqrt{10}}{10}$

c $5\sqrt{3}$ **d** $10\sqrt{6} - 5\sqrt{5}$

2 a $3\sqrt{3} + \sqrt{2}$ **b** $-5\sqrt{5}$

Part D

1 a false **b** true

c true **d** false

2 a I **b** R **c** R **d** I

Surds 2 PAGE 150

Part A

1 24 cm^2 **2** $\frac{3}{2}$

3 6 : 8 : 3 **4** $x^2 - 8x + 3$

5 $(7x - 3)^2$ **6** {HH, HT, TH, TT}

Part B

1 a $5\sqrt{11}$ **b** $28\sqrt{3}$

2 a $13\sqrt{2}$ **b** $5\sqrt{2} - 7\sqrt{5}$ **c** $23\sqrt{5} - 9\sqrt{7}$

Part C

1 a $15\sqrt{30}$ **b** 8 **c** -1

2 a $\frac{5\sqrt{2}}{2}$ **b** $\frac{2\sqrt{7}}{3}$ **c** $\frac{5\sqrt{3} - 3}{6}$

Part D

1a Multiply both the numerator and denominator by $\sqrt{5}$.

b The $\sqrt{5}$ in the denominator becomes rational (5) when multiplied by itself.

2 a $\sqrt{x} \times \sqrt{y}$ **b** $\frac{\sqrt{x}}{\sqrt{y}}$

c $a^2 - 2ab + b^2$

3 a $9 - 4\sqrt{5}$ **b** $-10 + 21\sqrt{2}$

CHAPTER 14

Quadratic equations PAGE 160

Part A

1 a 13 units **b** $-\frac{5}{12}$

2 a $(x + 11)(x - 8)$ **b** $5n(n + 5)(n - 5)$

3 a 13 **b** 6.5

Part B

1 a $x = -4, 6$ **b** $x = 0, 4\frac{1}{2}$

2 a $u = 3\frac{1}{2}, -1\frac{2}{3}$ **b** $k = 0, 3$

c $p = -2, \frac{3}{7}$

Part C

1 a $x = -3 \pm\sqrt{5}$ **b** $y = \frac{1 \pm \sqrt{3}}{2}$

2 a $a = -1 \pm \sqrt{6}$ **b** $b = 1, -3$

3 $x = \frac{-1 \pm \sqrt{85}}{6}$

Part D

1 25, 5

2 a $x = \frac{5 \pm \sqrt{10}}{2}$ **b** $x = \frac{2}{5}, -1$

c $x = 3, -\frac{1}{3}$

The parabola PAGE 162

Part A

1 a $(2a + 3)(3a + 1)$ **2** $x = 20y^2$

3 $m > 8$ **4** 12π cm^2

5 $x = 2\frac{2}{3}, y = 7$ **6** $2\sqrt{3}$

Part B

1 base 40 m, height 30 m **2** $x = -1, 2$

3 42, 44

Part C

1 a $x = 1$ **b** $(1, -4)$

2

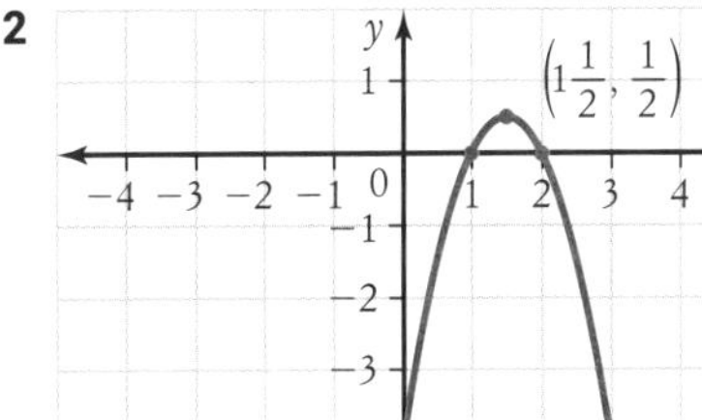

Part D

1 concave up, the coefficient of x^2 is $a = 1$, which is positive.

2 8 **3** $x = 3$

4 $(3, -1)$ **5** $x = 2, 4$

6

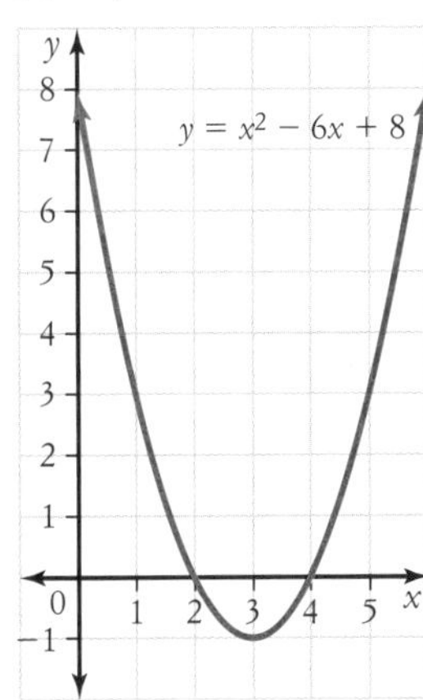

CHAPTER 15

Further trigonometry 1

PAGE 169

Part A

1 $19\sqrt{2}$ **2** \$6500 **3** $-\frac{1}{5}$

4 a $\frac{m^2}{4n^4}$ **b** $\frac{1}{27x^3}$

5 $\frac{3}{8}$ **6** $x = 5, -\frac{1}{2}$

Part B

1 a 20 cm **b** 22.4 cm **c** 27°

2 a 42° **b** 135 km **c** 318°

3 a $\sqrt{2}$ **b** 45°

Part C

1 −0.40

2 a 45° **b** 80° **c** 27°

3 $\theta = 48°, 132°$ **4** $x = 96°54'$

Part D

1 a negative **b** 0

c 1 **d** 60°, 300°

2 a $\sqrt{3}$ units **b** $x = 60, y = 30$ **c** $\frac{\sqrt{3}}{2}$

Further trigonometry 2

PAGE 172

Part A

1 $\frac{1}{25}$ **2** $\frac{9\sqrt{2}}{2}$

3 $(100b^2 + 9a^2)(10b + 3a)(10b - 3a)$

4 $x + 5y - 21 = 0$ or $y = \frac{-x+21}{5}$

5 a −8 **b** −2 and 2

Part B

1 a

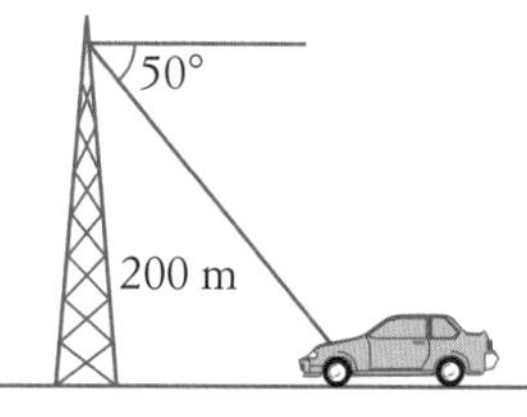

b 168 m

2 a west **b** southeast

3 12.4 km

Part C

1 a 4.23 **b** 51.20

2 a 54°20′ **b** 78°15′

Part D

1 cosine rule

2 $\cos C = \frac{a^2 + b^2 - c^2}{2ab}$

3 opposite, equal

4 122° **5** 426 km

Further trigonometry 3

PAGE 174

Part A

1 a 6.5 **b** 2

2 $\frac{x+2}{2x-5}$ **3** 150π m^2

4

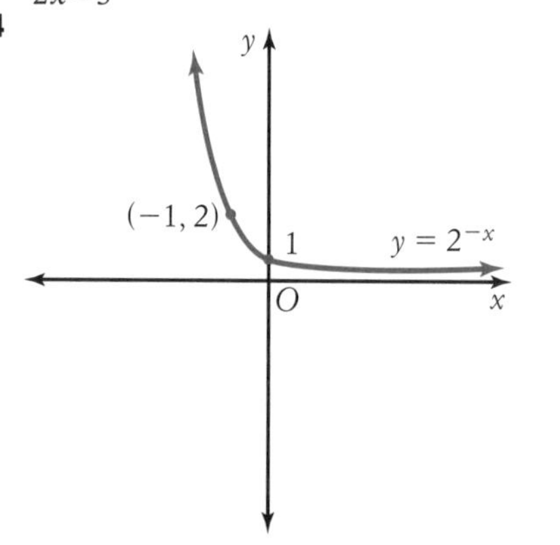

Part B

1 $x = 89$

2 a 18° **b** 64.96 m **c** 56 m

3 $x = 6.2$

Part C

1 a 183.2 cm^2 **b** 443.4 m^2 **c** 198.9 cm^2

2 a 1197.3 cm^2 **b** 567.9 cm^2 **c** 629.4 cm^2

Part D

1 included, sin C **2** sine rule

3 Pythagoras' theorem

4 a 1.5 km^2 **b** 8.5 km

9780170454551